Praise for **UNTHINKABLE**

AN AMAZON BEST NONFICTION BOOK OF THE MONTH

"Helen Thompson's remarkable book is an astonishing tour of the human brain in all its awesome power and bewildering variation. In beautiful prose, she seamlessly dances between conversations with nine extraordinary people and beautiful explanations of how the brain works. *Unthinkable* will enrich your brain, blow your mind, and warm your heart."

—Ed Yong, author of *I Contain Multitudes*

"An engaging tour inside the head. . . . A great science writer knows what is interesting to the reader, and here Thomson shines. Her book is tailor-made for anyone who loves intellectual brain trivia: why extroverts need more anesthesia, the advantage in wearing red on a date, the murdered priest behind the first lie detector. . . . This book is a chef's tasting menu of fascinating things about your brain—and a good one at that."

—*Washington Post*

"May change your perception of what it means to be human. . . . Thomson has a gift for making the complex and strange understandable and relatable."

—*Library Journal* (starred review)

"A window into neurological research that's not just about other people's brains, but also our own."   —*Popular Science*

"A stirring scientific journey, a celebration of human diversity and a call to rethink the 'unthinkable.'"   —*Nature*

"With a scientist's boundless curiosity and a writer's keen observation, Thomson imparts caring and humanity to each profile of these remarkable people. *Unthinkable* could easily sensationalize the weird and pervert the odd. Instead, Thomson underscores our commonalities and reminds readers that we all have truly extraordinary brains."   —*Booklist*

"We are the sum of our brains—nothing more or less. Helen Thomson ably guides us through the fascinating world of what are indeed some of the strangest brains on earth, showing us what we can learn about ourselves. Scientifically accurate and wholly accessible, this is an irresistible book."

—Robert Sapolsky, author of *Behave: The Biology of Humans at Our Best and Worst*

"With flair and empathy, the author sees her subjects in the context of their everyday lives, allowing us to marvel at their humanity. . . . This is neuroscience for the general reader: accessible, well researched, thought-provoking. Like Oliver Sacks's *The Man Who Mistook His Wife for a Hat*, Thomson's *Unthinkable* offers us a wondrous array of rare and weird disorders."

—*Tatler*

"Exceptional. . . . From seeing auras and turning into a tiger, to waking up 'dead' and being able to remember every single day of your life in vivid detail, award-winning science journalist Thomson investigates wondrously rare and strange brain disorders in this terrific debut. While acknowledging her debt to the late, great Oliver Sacks, to whose *The Man Who Mistook His Wife for a Hat* this bears more than a passing resemblance, Thomson sets out to do things a bit differently by meeting her nine subjects not in clinical environments but as they live their daily lives with extraordinary brains. Theirs are mystery stories, spellbinding and true."

—*The Bookseller*, Editor's Choice

"This wonderfully clear, fluent, eye-opening book explores what happens when the mind misbehaves: distance is distorted, memory plays tricks, people hear in colour and see in music. Helen Thomson is the science teacher you wish you'd had at school. . . . Fascinating."   —*The Times* (London)

"Refreshingly personal . . . humane and often humorous."

—*Evening Standard*

"Fun facts are what make popular science popular. Helen Thomson's first book has a ready supply of them, and she is good at giving them context. . . . There is much of interest here."   —*Sunday Telegraph*

UNTHINKABLE

UNTHINKABLE

An Extraordinary Journey Through
the World's Strangest Brains

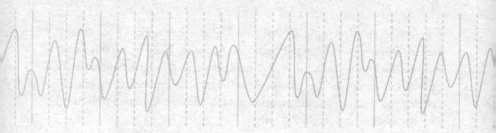

HELEN THOMSON

An Imprint of HarperCollinsPublishers

HarperCollins books may be purchased for educational, business, or sales promotional use. For information, please email the Special Markets Department at SPsales@harpercollins.com.

First published in Great Britain in 2018 by John Murray (Publishers) A Hachette UK company.

A hardcover edition of this book was published in 2018 by Ecco, an imprint of HarperCollins Publishers.

FIRST ECCO PAPERBACK EDITION PUBLISHED 2019.

Designed by Michelle Crowe

Library of Congress Cataloging-in-Publication Data has been applied for.

ISBN 978-0-06-239117-9

19 20 21 22 23 LSC 10 9 8 7 6 5 4 3 2 1

For Mum

CONTENTS

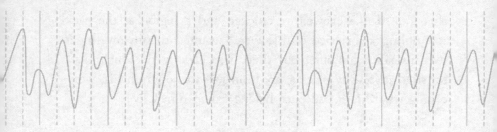

Introduction

THE STRANGE LIFE OF THE BRAIN

It's not something you easily forget, the first time you see a human head sitting upon a table. The worst part is the smell. The unforgettable stench of formaldehyde, the chemical fixative in which bits of the body are hardened and preserved. It gets up into your nostrils and really sticks around.

It wasn't the only head in the room; there were six, all severed at slightly different angles. This particular head had been lopped off just below the chin, and then sliced in half down the center of its face. It had belonged to an elderly gentleman—the deep wrinkles etched into his forehead held whispers of a long life. As I slowly circled the table, I saw a few gray hairs sticking out of a generous nose, an unruly eyebrow and a tiny purple bruise just above the cheekbone. And suddenly there it was, sitting in the middle of a thick bony skull—a human brain.

It had a grayish yellow tinge and a texture that conjured up thoughts of a shiny panna cotta. The outermost layer swirled around like a walnut. There were lumps and hollows, strands that looked like chewed-up chicken and a region at the back resembling a shriveled cauliflower. I wanted to run my finger over its silky contours, but there was strictly no touching. I satisfied myself by placing my head just inches from his, wondering what life it had once held. I called him Clive.

I HAVE ALWAYS BEEN interested in people's lives. Perhaps this is why I was compelled to study the human brain at university. The two are, after all, inextricably linked. Everything that we feel, every story we experience or tell, we owe to that three-pound lump of mush in our heads.

That may seem obvious today, but it wasn't always so clear. The first mention of the brain was by the ancient Egyptians in a surgical scroll called the Edwin Smith Papyrus. They wrote that the way to identify the brain was to "probe a head wound and see whether it throbs and flutters under your fingertips."[1] But seemingly it was an organ of little interest. If a head wound had occurred, they would pour oil on it and take the patient's pulse "to measure his heart . . . in order to learn the knowledge that comes from it." For it was the heart, not the brain, that was believed to house our mind at the time. After death, the heart was carefully preserved inside the body to allow safe passage into the afterlife, while the brain was fished out piece by piece through the nose.

It was only around 300 BC, when Plato began grappling with the idea that the brain was the seat of the immortal human soul, that it gained a greater significance in medical thought. But although his teachings would later influence many scholars, his own peers were not convinced. Even Plato's best student, Aristotle, continued to argue that the mind was contained in the heart. Physicians at the time were reluctant to open human cadavers, fearful of preventing their owners' souls from reaching the afterlife. So his arguments were largely based on dissections of the animal kingdom, which revealed that many creatures had no visible brain at all. How then could it have any vital role?

Aristotle declared that the heart carried out the responsibilities of the rational soul, providing life to the rest of the body. The brain was there simply as a cooling system, tempering "the heat and seething" of the heart.[2]

(Later we'll find out that they may both have been correct—that you cannot think or feel without both your heart and brain communicating with each other.)

The Greek anatomists Herophilus and Erasistratus finally got the chance to dissect the human brain in 322 BC. Less concerned with identifying the soul, they concentrated on basic physiology, discovering the network of fibers that run from the brain to the spine and out around the body—what we now refer to as the nervous system.

It was down in the gladiator stadium, however, where the brain really came into its own. The philosopher, physician and writer Claudius Galen was forbidden by Roman law from dissecting the human brain for himself, so instead he would head to the dusty arena, where he could gain a snapshot of the brain's anatomy by treating bloodied soldiers whose skulls had been torn apart in combat.

But it was his experiments on live, squealing pigs that caused the biggest sensation. In front of a large crowd, he would slice through the laryngeal nerve connecting the pig's voice box to its brain, and watch as the pig fell silent. The crowds would gasp—Galen had offered the first public demonstration that the mind, not the heart, controlled behavior.

Galen also discovered four cavities within the human brain, later called ventricles. We now know that ventricles are spaces containing fluid that protects the brain against physical knocks and disease, but Galen's prevailing view was that all aspects of the immortal soul floated around these ventricles. It then passed into "animal spirits," which were pumped around the body. This explanation particularly suited those high up in the Christian church, who were growing increasingly concerned about the idea that the brain could provide a physical basis for the soul. How could something be immortal if it was present in such frail flesh? It was much more palatable to place our soul in these "empty" spaces instead.

GALEN'S THEORIES OF THE BRAIN reigned for fifteen centuries, and religion continued to influence those who built upon his ideas. René Descartes, for instance, famously declared that the mind and the body were separate—what is now known as dualism. The mind was immaterial and did not follow the laws of physics. Instead, he said, it did its bidding via the pineal gland, a small pine-nut-shaped region in the center of the brain. The pineal gland would move, letting out the particular animal spirit required to carry out the soul's needs. His purpose in showing this distinction was to rebut those "irreligious people" who would not believe in the soul's immortality without a scientific demonstration of it.

But it was in the dirty, smoke-filled streets of seventeenth-century Oxford where things really started to get interesting. Down in the bowels of the city's university, a resourceful young physician called Thomas Willis was sharpening his scalpel.

In front of a large audience of anatomists, philosophers and interested public, he would carve up the human body and brain, demonstrating its intricate anatomy to anyone who cared to watch. He had been given permission to do so by King Charles I, who allowed him to dissect any criminal sentenced to death within the city. It was thanks to this that he created meticulous illustrations of the human brain, and was said to have become "addicted . . . to the opening of heads."[3]

I mention Willis for it was he who really began to cement the idea that our human identity was connected to the brain.

He started to match the altered behavior he observed during his patients' lives to deformities he discovered during autopsy. For instance, he noted that people who had pains in the back of their head, near to an area of the brain called the cerebellum, also had pain in their heart. To prove that the two were linked, Willis opened up a live dog and clamped the nerves running between the two—the dog's heart stopped and the animal died almost immediately. Willis went on to examine how the brain's chemistry might produce other aspects of our lives: dreams, imagination and memories. It was a project that he called "neurologie."

In the nineteenth century, the German anatomist Franz Joseph Gall pulled us closer still to our modern understanding of the brain by advocating the idea of localization. The brain, he said, was comprised of specific compartments, each responsible for a fundamental faculty or tendency, including a talent for poetry and an instinct for murder. He also thought that the shape of the skull could determine personality. Gall had a friend who had big bulging eyes, and because his friend also had a fantastic memory and was great with languages, he believed that the brain regions responsible for these abilities must be located behind the eyes, and had grown so large that they were pushing the eyeballs outward. Despite phrenology later being discredited, Gall's idea of the brain being made up of discrete regions was prescient—in some cases he was even correct in pinpointing their responsibilities. His "organ of mirthfulness," for instance, was placed toward the front of the head, just above the eyes. In later years, neurologists would come to stimulate this area and in doing so make a patient burst out laughing.

Gall's observations ushered in a new age of the brain—one that separated itself from the philosophy-driven science of prior centuries. Later, the acceptance of atoms and electricity allowed us finally to bid farewell to the animal spirits of the past. Nerves were no longer hollow conduits through which the soul's desires were driven, but cells that crackled with electrical activity.

Although scientists in the nineteenth century focused on using electrical stimulation to identify which bits of the brain carried out what functions (no doubt spurred on by the fact that they got to name the regions after themselves), those of the mid- to late-twentieth century placed more emphasis on the ways in which these areas communicate with each other. They discovered that communication between different regions of the brain was more important in bringing about complex behavior than the action of any one region alone. Functional MRI, EEG and CAT scans allowed us to view the brain in intimate detail, even examine its activity while hard at work.

Through these tools, we now know that there are 180 distinct regions that lie within that three-pound lump of tissue that throbs and flutters within our skulls. And back in the anatomy room at the University of Bristol, I was tasked with gaining an intimate knowledge of each one.

AS I STARED AT CLIVE, I could easily spot the most recognizable region of the human brain—the cerebral cortex. This forms the outside shell and is divided into two almost identical hemispheres. We tend to carve up each side of the cortex into four lobes, which together are responsible for all our most impressive mental functions. If you touch your forehead, the lobe closest to your finger is called the frontal cortex and it allows us to make decisions, controls our emotions and helps us understand the actions of others. It gives us all sorts of aspects of our personality: our ambition, our foresight and our moral standards. If you were then to trace your finger around either side of your head toward your ear, you would find the temporal lobe, which helps us understand the meaning of words and speech and gives us the ability to recognize people's faces. Run your finger up toward the crown of your head and you'll reach the parietal lobe, which is involved in many of our senses, as well as certain aspects of language. Low down toward the nape of the neck is the occipital lobe, whose primary concern is vision.

Hanging off the back of the brain we have a second "little brain," that distinctive cauliflower-shaped mass. This is called the cerebellum and it is vital for our balance, movement and posture. Finally, if you were to gently pry open the two hemispheres (a bit like pulling apart a peach to reveal the stone), you would find the brain stem, the area that controls each breath and every heartbeat, as well as the thalamus, which acts as a grand central station, relaying information back and forth between all the other regions.

Although they are too tiny to see with the naked eye, the brain is full of cells called neurons. These cells act like wires from an old-fashioned telephone system, passing messages from one side of the brain to the other in the form of electrical impulses. Neurons branch out like twigs on a tree, each forming connections with its neighbors. There are so many of these connections that if you were to count one every second, it would take you three million years to finish.

We now know that the mind arises from the precise physical state of these neurons at any one moment. It is from this chaotic activity that our emotions appear, our personalities are formed and our imaginations are stirred. It is arguably one of the most impressive and complex phenomena known to man.

So it's not surprising that sometimes it all goes wrong.

JACK AND BEVERLY WILGUS, vintage-photography enthusiasts, don't recall how they came by the nineteenth-century image of a handsome yet disfigured man. They called him "the Whaler" because they thought the pole he held in his hand was part of a harpoon. His left eye was closed, so they invented an encounter with an angry whale that left him with one eye stitched shut. Later, they discovered that it wasn't a harpoon but an iron rod, and that the photo was the only one known of a man called Phineas Gage.

In 1848, the twenty-five-year-old Gage was working on a railroad bed when he was distracted by some activity behind him. As he turned his head, the large rod he was using to pack powder explosives struck a rock, caused a spark and the powder exploded. The rod flew up through his jaw, traveled behind his eye, made its way through the left-hand side of his brain and shot out the other side. Despite his somewhat miraculous survival, Gage was never the same again. The once jovial, kind young man became aggressive, rude and prone to swearing at the most inappropriate times.

As a toddler, Alonzo Clemons also suffered a traumatic head injury, after falling onto the bathroom floor. Left with severe learning difficulties and a low IQ, he was unable to read or write. Yet from that day on he showed an incredible ability to sculpt. He would use whatever materials he could get his hands on—Play-Doh, soap, tar—to mold a perfect image of any animal after the briefest of glances. His condition was diagnosed as acquired savant syndrome, a rare and complex disorder in which damage to the brain appears to increase people's talent for art, memory or music.

SM, as she is known to the scientific world, has been held at gunpoint and twice threatened with a knife. Yet she has never experienced an ounce of fear. In fact, she is physically incapable of such emotion. An unusual condition called Urbach-Wiethe disease has slowly calcified her amygdalae, two almond-shaped structures deep in the center of the brain that are responsible for the human fear response. Without fear, her innate curiosity sees her approach poisonous spiders without a second's thought. She talks to muggers with little regard for her own safety. When she comes across deadly snakes in her garden, she picks them up and throws them away.

By the end of my degree, it had become clear to me that unfortunate accidents, maverick surgeries, disease and genetic mutations are often the reason we discover how different bits of the brain work. Gage showed us that our personalities were intimately tied up in the front regions of the brain. Studies on autistic savants like Clemons have propelled our understanding of creativity. Even today, scientists continue to try to scare SM, in the hope that they'll have a better understanding of how to treat those who fear too much. I was enchanted by this concept: The strangest, most unique brains are often those that teach us the most about our own.

OF COURSE, NOT SO LONG AGO, having an unusual brain would have seen you carted off to an asylum. "Mental illness" is a term that has been in use only for the past two hundred years; prior to that, any strange behavior would have been considered madness, and blamed on anything from curses to demons to an imbalance of humors in the body.[4] If you lived in England and were suffering from such madness, you might have found yourself in Bethlem Hospital, popularly known as Bedlam. In his book *This Way Madness Lies*, Mike Jay refers to Bethlem as the stereotypical eighteenth-century madhouse, later a nineteenth-century lunatic asylum, and now a model example of a twenty-first-century psychiatric hospital.[5]

The different incarnations of the hospital reflect how society has undergone a radical transformation in its treatment of the strange brain. When Bethlem was first founded, it specialized in keeping off the street those referred to as "lunaticke." Its guests were violent or delusional, had lost their memory, speech or reason. They were locked up among vagrants, beggars and petty criminals.

Patients were given general treatments aimed at restoring a healthy constitution. These included bloodletting, cold showers and emetics that made them vomit up anything that might be blocking their digestion. It was the madness of King George III that prompted a shift in this attitude. George had been taken ill with a stomach bug but soon started foaming at the mouth and showing signs of insanity. Clergyman Francis Willis was called; he had a formidable reputation for curing such illness. His approach was straightforward: he put George to work in the fields, dressed him well, made him exercise and encouraged "good cheer." Over three months George's mental health improved alongside his physical symptoms. The idea that madness was something that could be corrected began to percolate within the medical community. Through the nineteenth century, asylums progressed alongside increasingly rational explanations of how the mind worked. There were still a few bumps in the road—straitjackets were a common sight and many therapies would be considered barbaric by today's standards—but doctors also began to think about how the wider family might help their patients, how interaction with the outside world could be established and what drugs might help ease pain and subdue anxieties. In the early twentieth century "insanity" was rebranded "mental disease," and physicians began to conceive of a biological basis for disorders of the mind. Just as Thomas Willis predicted, they were able to look into the brain and start to pinpoint the exact changes that correspond to unusual behaviors and perceptions.

Today we understand that mental illness, or in fact any mental anomaly, can be the result of small malfunctions in electrical activity, hormonal imbalances, lesions, tumors or genetic mutations—some of which we can fix, some we can't, and some that we no longer see as a problem.

We are by no means close to understanding the mind in its

entirety. In fact, none of what we call our "higher" functions—memories, decision-making, creativity, consciousness—are close to having a satisfying explanation. For instance, we can spark a hallucination in anyone using a simple ping-pong ball (I'll show you how later), yet we have few ways to treat the hallucinations that characterize schizophrenia.

What is clear is that the strange brain provides a unique window into the mysteries of the so-called normal one. It reveals some of the extraordinary talents locked up inside us all, waiting to be unleashed. It shows us that our perceptions of the world aren't always the same. It even forces us to question whether our own brain is as normal as it would have us believe.

AFTER COMPLETING MY DEGREE in neuroscience, I decided to become a science journalist. I figured it was the best way to discover new and mysterious ways in which the brain works while simultaneously feeding my passion for learning about people's lives and telling a good story. I studied for a master's in science communication at Imperial College London and then worked my way up to becoming a news editor at *New Scientist* magazine.

Now, as a freelance journalist, I work for a variety of media outlets, including the BBC and the *Guardian*. But despite writing about all sorts of health matters, I always find myself being drawn back toward the strange brain. I attend neurological conferences, devour scientific papers and collect stacks of quirky medical journals for the merest hint of a study that describes someone with an unusual mind. Nothing else fascinates me half as much.

It's not an easy job. Gone are the case studies of old—the rich tales presented by the case-historians of the eighteenth and nineteenth centuries, who would describe their patients' lives in full and glorious color. Today's case studies are objective, cold and impersonal. Patients are known only by their initials, their defining characteristics are lost, their lives go unmentioned. The subject of neurology—the *owner* of the brain in question—has largely become inconsequential to the science that surrounds them.

But one evening, late in the office, I came across a paper unlike any other. It described a condition, first discovered in 1878, deep in the forests of Maine. There was a mysterious behavior afflicting a small group of lumberjacks, and the American neurologist George Miller Beard had been asked to investigate. What he found seemed implausible. Among this group were a few men whom Beard later called "the Jumping Frenchmen of Maine." Startle a jumper with a short, verbal command and he will obey it and repeat the command, immediately, no matter the consequence. Tell him to throw a knife and he will throw it. Tell him to dance, and he will dance.

What stood out as much as the description of the disorder itself was the picture on the second page. It was of a woman who had the condition. There she was, leg in the air, mid-startle. It was taken in her own home. It was the first time I had seen a photograph of a case study published in a scientific paper in years.

Beard had spent weeks in those woods, and in the hotels where the jumpers worked in the off-season. He had spoken to their friends and families. He had written about their hobbies, their relationships. He had tried to find out about their brains by learning about their lives. He told a fascinating story.

Staring at the picture, I wondered what would happen if I did the same thing today. Could I follow in Beard's footsteps and find out about the most unusual aspects of the human brain by going out and meeting the people who live with them?

I was reminded of something Oliver Sacks once said: To truly understand someone, to get any hint of one's depth, you need to lay aside the urge to test and get to know your subject openly, quietly, as they live and think and pursue their own life. There, he said, is where you will find something exceedingly mysterious at work.

I glanced at the pile of papers stacked in front of me—a ten-year collection of the strangest brain conditions known to science, most of them described only by their initials, their age and their sex. I carefully lifted the pile from the desk and spread the pieces out on the floor around me. I sat there reading for hours. All over the world strange things were happening to normal people—what kind of lives did they lead? I wondered. And would they let me tell their stories?

OVER THE NEXT TWO YEARS, I traveled around the world to meet people with the most extraordinary brains. They have all been tested, scanned and analyzed by multiple doctors and researchers, but they have rarely—if ever—publicly divulged information about their lives. Sacks, of course, did something similar on a number of occasions, most notably in his 1985 book *The Man Who Mistook His Wife for a Hat*. In this book he calls his case studies "travellers to unimaginable lands."[6] Without their stories, he says, we would never know that such perceptions of the world were possible.

I felt it might be the right time to revisit this idea, to see what a thirty-year neurological revolution had revealed. What new lands might have appeared? I also wanted to do something Sacks hadn't. I wanted to divorce these case studies completely from the hospital environment and from the eyes of a neurologist. I wanted to see them as a friend might, play a part in their world. I wanted to ask the questions that scientists avoid. I wanted to hear stories of their childhood, how they find love, how they navigate the world when their mind works like no other. I wanted to understand how their life differs from my own.

I wanted to know just how extraordinary the brain could get.

I started my journey in America, where I met a TV producer who never forgets a day of his life, and a woman who is permanently lost—even in her own home. In the UK, I spent time with a teacher whose memories don't feel like her own, and the family of an ex-con whose personality changed overnight. I flew across Europe and the Middle East to meet with a man who turns into a tiger, a woman who lives with a permanent hallucination, and a young journalist who sees colors that don't exist in real life. And then there was Graham, a man who, for three years, believed he was dead.

I met with people who had embraced their strange brain for years, and others who had kept it secret from the world until now. Along the way, I came across researchers on the fringes of science, people trying to answer questions about the nature of reality, about the existence of auras, about the limits of human memory. And toward the end of my travels I met a man, a doctor no less, whose brain was so extraordinary it changed the way I felt about what it means to be human.

At the beginning of my journey I wondered whether I would be able to understand their unique encounters with the world. What I discovered was that, by putting their lives side by side, I was able to create a picture of how the brain functions in us all. Through their stories I uncovered the mysterious manner in which the brain can shape our lives in unexpected—and, in some cases, brilliant and alarming—ways. But they also showed me how to forge memories that never disappear, how to avoid getting lost, and what it feels like to die. They taught me how to make myself happier in a split second, how to hallucinate, how to make better decisions. I learned how to grow an alien limb, how to see more of my reality, even how to confirm that I am alive.

I can't say exactly when it happened. Maybe it was the time

I started seeing people who didn't exist, or the moment I discovered the way to hear the sound of my own eyeballs moving. But somewhere between a blizzard in Boston and a dusty camel racetrack in Abu Dhabi, I came to the realization that I wasn't just learning about the most extraordinary brains in the world, but was uncovering the secrets of my own.

Some of my tales begin very recently, others centuries ago. And so it is that we start this journey not in the twenty-first century but back in ancient Greece, at a banquet, moments before a terrible catastrophe is about to occur.

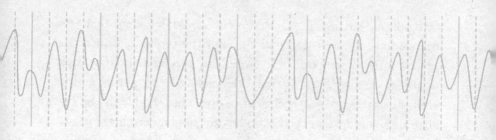

BOB

Never Forgetting a Moment

In 500 BC, the poet Simonides of Ceos was sitting in a large banquet hall. Rather than enjoying the meal, he was angry with his host—Scopas, a rich nobleman. Simonides had been promised a fine sum of money to compose a poem in his host's honor, which he had recited to the guests. But Scopas had refused to pay him the full fee. He told Simonides that the poem spent too long referring to the mythical twins Castor and Pollux and not enough time on his recent victories.

Midway through his main course, Simonides received word that two young men were waiting outside for him. He left the building just in time; as he stepped through the door, the roof of the banquet hall came crashing down, killing everyone inside. The two young men were nowhere to be seen, prompting later rumors that they were in fact Castor and Pollux, who had saved Simonides's life as a reward for his belief in them.

As the dust settled and the rubble was removed, it became clear that the people inside the hall were crushed beyond all recognition, so disfigured that none could be identified. As friends and relatives searched through the remains, Simonides surveyed the destruction. He closed his eyes and thought back to where he had been sitting. He pictured the guests eating around him, Scopas at the head of the table. Suddenly, he realized he could identify the bodies by remembering the exact location in which everyone had been seated. In that moment, Simonides began to unlock the secrets of memory.

IT'S CROWDED, HOT AND NOISY at Heathrow Airport, where I'm sitting waiting to board my delayed twelve-hour flight. To pass the time, I've taken to watching two children play a game on the floor in front of me. One by one they turn over cards to reveal brightly colored animals. When they reveal two of the same kind they get to keep both cards. It seems appropriate, I think, as I mentally play along.

It wasn't a difficult decision, working out who to visit first. When looking back over all the extraordinary people I'd come across in my career as a science journalist, the one who immediately came to mind was Bob—a man whom medical papers describe as being able to remember every single day of his life.

I thought about Bob a lot.

I'd thought about Bob as I stared at the strange pile of food on the kitchen counter earlier that month. It was Sunday afternoon and I had sent my husband, Alex, out on an errand. I asked him to get some oranges, pasta and a bulb of garlic. Twenty minutes later he returned with three bananas, an onion and some dog food. I thought, not for the first time, what a strange thing memory was.

I'd thought about Bob when a week earlier, having arrived at work, I was suddenly convinced that I had left my kettle bubbling away on the gas stove. Over and over again I replayed the events of the morning, but just couldn't remember if I had turned the gas off. I imagined the steam pouring out of the kettle's spout. I saw the water boiling and evaporating until the flame started burning its dry base. By the time I returned home, I was convinced that I would see a smoldering pile of ash where my house once stood. Despite the relative calm at the front of the building, I rushed inside and into the kitchen, where the kettle rested quietly on the unlit stove.

I thought about Bob as I sat there watching the children turn the cards over and over.

I find it odd that something so fundamental to our daily lives so often fails us. Why is it that I can remember the first time I built a snowman, my seventh birthday cake, or the phone number of a friend I haven't seen for twenty years, while other memories, far more vital to my present well-being, float away as if they had never happened? How many hours of my life have I spent trying to remember things that I had forgotten? Where I put my keys, whether I'd fed the dog, when the trash went out, why I came downstairs. Sure, there were bits of my life that I'd happily forget forever, but there was so much more that I wished I could remember. It seemed like the obvious place to start my journey—to meet Bob and find out what it is like to have the perfect memory.

HAVE YOU EVER THOUGHT about what a memory actually is? Scientists have been searching for the answer to that question for centuries. In the 1950s, a piece of the puzzle arrived in the shape of Henry Molaison.

A handsome young child with dark sweeping hair and a strong jawline, Molaison had a promising life in front of him. But he noticed the cyclist speeding along the road just a second too late. It was never clear whether or not his seizures were caused by the accident, but by the time he reached twenty-seven they had become so bad that he was unable to work. In 1953, Molaison agreed to an experimental technique that had never been trialed before. In an attempt to cure his seizures, doctors drilled holes into his brain and sucked out the areas responsible for them—a seahorse-shaped region on either side of the brain called the hippocampus. The operation was a success in that it largely cured his seizures, but it had one disastrous side effect: Molaison could no longer form any conscious, long-term memories. Despite retaining a great deal of information that had occurred before the operation, he forgot any experience after the surgery within thirty seconds.

A young postgraduate researcher called Suzanne Corkin met with Molaison, and began to study him. In a book she later wrote about their friendship, she called him a willing student.[1] She said that in living in a world bound by only thirty seconds he was not stressed by the anxiety that stems from worries about the past or plans for the future. And as weeks turned into months, something unexpected began to happen.

It started when Corkin, and her former supervisor, Brenda Milner, at McGill University, showed Molaison a sketch of a five-pointed star.[2] They then asked him to trace its outline with a pencil but only by looking at his drawing hand and the star as reflected in a mirror. Try it for yourself: it's not an easy task. Over time, Molaison became better at this skill, and others like it, despite having had no recollection of having performed them before. It proved he could retain long-term memories for motor movements. His unique brain provided the first fundamental evidence that specific types of memory are processed in different places, and indicated where those memories might be stored. Corkin continued to meet with Molaison regularly over the next forty-six years, although for Molaison each day they spoke was like the first. "It's a funny thing," he told her. "You just live and learn. I'm living and you're learning."[3]

MORE THAN HALF A CENTURY after Molaison's surgery, scientists are still debating the exact nature of memory. Most agree that there are three kinds—sensory, short-term and long-term. Sensory memory is the very first kind of memory that enters your brain; it lasts for just a split second—just enough time for you to sense your environment. The touch of your clothes against your skin, the smell of a bonfire in the air, the sound of the traffic outside. But unless we attend to that memory, it disappears for good. Ten seconds ago you didn't notice your socks against your feet. It shot into your brain and right back out again. Now you can't stop thinking about your socks, and that's because I mentioned them, nudging that sensory memory into your short-term memory.

Short-term memory is your memory of current events—the things that you are thinking about right now. You use it all the time without realizing. For instance, you can only understand what happens at the end of this sentence because you remember what happened at the beginning. Our short-term memory is said to have a finite capacity of about seven items, which can be held in mind for fifteen to thirty seconds. Rehearse those items, however, and you could transfer them into your long-term memory—our seemingly limitless warehouse for storing recollections for the long haul.

This is arguably our most important type of memory. It is this that allows us to time-travel mentally to the past, and also to predict our future. It is no exaggeration to declare that memory allows us to make sense of the world. In his autobiography, the filmmaker Luis Buñuel sums it up neatly: "Life without memory is no life at all . . . our memory is our coherence, our reason, our feeling, even our action. Without it, we are nothing."[4]

Solomon Shereshevsky's editor was incredibly annoyed. He had just come out of a news meeting in which he had given Shereshevsky a large list of instructions—people he needed to interview, information about a breaking story, addresses of places he had to visit. As usual, Shereshevsky didn't take a single note. The editor was going to have to say something. He called him into his office, sat him down and told him off for being inattentive. Shereshevsky was unapologetic. He hadn't needed to take any notes, he said, and proceeded to repeat back his editor's complicated instructions word for word.

Shocked, Shereshevsky's editor persuaded him to pay a visit to Alexander Luria, a Russian psychologist. Luria discovered that the secret to Shereshevsky's perfect recall was a condition known as synesthesia. This is when a person experiences the joining of senses that are normally experienced apart. They might, for instance, taste lemon when they hear the sound of a bell, or see red when they think of a number. We will encounter this condition several times in this book. Shereshevsky's linked senses meant that if asked to memorize a word, he would also taste and hear that word simultaneously. This meant that when recalling the word at a later date, he had several triggers to remind him of it. Shereshevsky's imagination was so vivid that in one experiment, he was able to raise the temperature of one hand while lowering the temperature of the other, merely by imagining one on a stove and one on a block of ice.

Luria began to test Shereshevsky in the 1920s, and continued to do so for thirty years. According to his notes, he eventually gave up trying to find the limits of his incredible memory.[5]

While there are very few accounts of people with such natural talents of recall, there are many about those who have learned to perform extraordinary feats of memory. Take George Koltanowski, for instance, who took up chess at the age of fourteen, and three years later was Belgian champion. He was also able to play blindfolded by memorizing his opponents' moves after being told them by a referee. In 1937, he set a world record by playing thirty-four simultaneous games of chess blindfolded. His opponents were sighted, yet he won twenty-four games and drew ten. His record remains unbeaten today.

While impressive, Koltanowski did not have a memory that was naturally superior to yours or mine. Instead, he learned ancient parlor tricks such as mnemonics, which help you to associate information you want to learn with something more fun and memorable, like a funny image, or a rhyme or ditty.

Which is why, when I first heard about Bob—a man who could remember every day of his life—I assumed he must be doing something similar. But something didn't add up. Surely there wasn't enough time in the day to remember everything that has happened using rhymes and ditties? I looked back through the medical literature to find mention of anyone who had his talents, and discovered that until very recently the idea of having a perfect memory of your own past was unheard of.

That was until the American neurobiologist James McGaugh received a very odd email.

In 2001, McGaugh was puttering around his office when his computer *pinged*. It was an email from a woman who had googled "memory" and had hit upon his name. The woman, who was later revealed as Jill Price, a school administrator from California, told McGaugh that she had a strange memory problem and that she'd like to meet with him. McGaugh, an expert in the field of learning and memory, but no longer practicing medicine, responded simply by referring her to a specialist memory clinic. Jill replied immediately. She said, "No, I'd like to talk to you—because I don't forget. Anything."

I just hope somehow you can help me. I am thirty-four years old and since I was eleven I have had this unbelievable ability to recall my past, but not just recollections . . . I can take a date, between 1974 and today, and tell you what day it falls on, what I was doing that day and if anything of great importance occurred on that day . . . Whenever I see a date flash on the television I automatically go back to that day and remember where I was, what I was doing, what day it fell on and on and on and on and on. It is nonstop, uncontrollable and totally exhausting.[6]

Jill went to McGaugh's lab on a Saturday morning that spring. He took a large book from his shelf and opened it at random. It was a present he had been given that Christmas, and it contained newspaper clippings from every day of the past century. McGaugh randomly chose a date from Jill's lifetime and asked her what had happened on that day.

"She did an incredible job," says McGaugh, recollecting that first meeting for me. "I'd give her an event and she'd tell me which day and date it happened on, or I'd give her the date and she'd give me the event."

McGaugh also asked her to give the dates of the last twenty-one Easters, which she did without error. She even told him what she did on each of those days, an even more remarkable feat, given that she was Jewish.

Was it just a trick? Had Jill figured out how to apply the mind games that served Koltanowski so well to remembering periods of her own life? To find out, I decided to learn a few tricks of my own.

If you'd told Alex Mullen a few years ago that he was capable of remembering a whole pack of cards in less time than it takes to tie a pair of shoelaces, he would have said you were being ridiculous. His memory wasn't anything special, "below average" even.

"What on earth happened?" I asked him.

"I read this book," he said. "It was called *Moonwalking with Einstein*."

The book was written by Joshua Foer, a journalist who had attended a U.S. memory championship to write about what he thought would be "the Super Bowl of savants."[7] Instead, he found a group of people who had trained their memory using ancient techniques. Foer started practicing the techniques, and went on to win the competition the following year.

Mullen, an American medical student, was spurred on by Foer's story. He too started practicing. Two years later, he found himself in Guangzhou, China, in second place in the final round of the 2015 World Memory Championship. The competition consists of ten rounds of mental challenges, which include memorising as many numbers as you can in an hour, remembering as many faces and names as you can in fifteen minutes, or committing to memory hundreds of binary digits. The final event is always the speed-card round, where competitors memorize the order of a shuffled pack of cards as fast as possible— it was one of Mullen's favorites. That day, Mullen memorized all fifty-two in 21.5 seconds—one second faster than Yan Yang, the current competition leader—earning him just enough points to creep into first place and to win the entire championship.

These feats of memory may seem outrageous. But according to Mullen, anyone can do it. "You just have to create a mind palace," he said.

For those of you who aren't familiar with Sherlock Holmes,

a mind palace is an image in your mind's eye of a physical location that you know well. Perhaps it's your house or your route to work. To remember many items, be it cards or groceries, you just walk through your mind palace and drop off an image of each item at specific places along the way. To recall the items you merely have to retrace your steps and pick them up again.

This was the technique developed by Simonides of Ceos after the banquet ceiling fell to the floor. His ability to identify the bodies, based on remembering where they had been sitting, led him to discover that the best way of remembering anything is to attach an image of it to a familiar and orderly location.

Try it now with some of the things around you. Since I'm sitting at my desk at home, I think about memorizing my stapler, a cup of tea, my printer, my notepad and so on. My natural mind palace is my route to work. So I give the stapler to the woman at my local petrol station, who, in my imagination, uses it to staple my receipt together. I leave my cup of tea at the bus stop, placing it underneath the seats so it doesn't get spilled. I lug my printer all the way to the station, where I leave it with the ticket seller, before getting on the train and wedging my notebook between two seats. Not only should you be able to remember your items in the order you dropped them off, you should also be able to travel backward and name them in reverse.

If you want to remember large groups of numbers, though, you're going to have to learn another trick. Our memories haven't evolved to store all types of information equally well. Experiences that are more important to our survival stick around more easily than those that are less essential—and numbers, not being vital to our immediate well-being, are low down on the list. To get around this problem, we have to convert such information into visual imagery—pictures that our memories prefer to store. To remember a whole set of playing cards, Jonas von Essen, a student at the University of Gothenburg, and a former world memory champion, told me that he links each numbered card with an image. He then groups these images into sets of three before placing them around his mind palace. For him, the four of hearts, the nine of hearts and the eight of clubs instantly transform into an image of Sherlock Holmes playing the guitar while eating a hamburger.

As soon as von Essen tried the technique, he realized he could "memorise more things than I ever dreamed possible." Next year, he's hoping to break the world record for memorizing pi—his aim is to reach one hundred thousand digits.

Was it really that simple? I wondered. Could anyone use this technique to become a memory champion? Or was there more to it? Researchers at University College London wanted to know the answer, so they scanned the brains of ten people who had placed at the highest levels of the World Memory Championship. As is usual for such tests, they also scanned the brains of similar-aged people who had regular memory skills. By taking a look inside their brains, they hoped to identify whether the super-memorizers had any structural brain differences that predisposed them to having such an extraordinary talent.

As expected, when asked to memorize sets of three-digit numbers, the super-memorizers performed much better than the control group. But when it came to memorizing close-up images of snowflakes, neither group did very well. When I asked Eleanor Maguire, the lead researcher on the study, what they had discovered, she said that their tests could not establish any difference in intellect, nor any structural anomalies in the brain. But there was one vital difference between the two groups: while recalling sets of numbers, the memory champions appeared preferentially to use three brain regions that are associated with spatial awareness and navigation.[8] In other words, she said, the super-memorizers were better at remembering purely because they were walking around their mind palaces.

"Does it work every time?" I asked von Essen. "Doesn't your memory ever go blank?"

"No," he said. "If you've put it in your mind palace, it's always safe."

THE PROBLEM FOR McGAUGH was that Jill didn't seem to be using any of these parlor tricks. Time and time again Jill said that her memory was automatic and not strategic. Her memories came to her like images in a movie, full of emotion, under no conscious control. McGaugh believed her, noting that her answers to his questions were "immediate and quick, not deliberate and reflective."

McGaugh spent the next five years working out just how special Jill's memory was. Luckily, she had written detailed diaries from the age of ten to the age of thirty-four, which allowed his team to verify her account of thousands of personal events.

What became clear was that, despite her unprecedented memory for personal experiences, Jill wasn't that great at any other kind of memory task. She couldn't remember strings of numbers or items on a table any better than others her age. She hadn't excelled at school; she said she found it hard to remember facts and figures. It was unexpected: Jill didn't have a photographic memory—she had an exceptional autobiographical memory.

McGaugh wondered why Jill's memory of the events of her own past was so vivid while the other types of memory were mundane. As far as he was aware, there was no one else like her, and barely any scientific literature about such a superior form of memory. It was like a detective story, he said. To find more clues he needed more evidence, and that meant more people. So he published a paper on Jill, and named the condition "highly superior autobiographical memory" (HSAM).[9] His paper got picked up by the international press, and he was inundated with people claiming to have a similar talent. He and his colleagues began the long process of testing them all. Only five people passed their rigorous examinations. One of those was Bob.

"Sorry I'm late," says Bob. "I forgot where this place was."

It is early evening in L.A., I'm severely jet-lagged and haven't even dropped my suitcase off at my hotel. I laugh uneasily.

Bob and I are at Truxton's American Bistro in Westchester. We take a seat at the bar and order a glass of beer. Bob, a sixty-four-year-old TV producer with thin black glasses, has a crooked smile and a slightly nasal voice that reminds me of a cartoon.

As it turns out, he wasn't joking—Bob actually had forgotten where the restaurant was. Like Jill, having a great memory for his own past didn't make him any better at remembering other kinds of facts. But ask him to remember any day in his life and it's a completely different story. He can remember a day from forty years ago as easily as a memory of yesterday. The day comes flooding back as a rich, multisensory experience, complete with smells, tastes and emotions.

"It's like watching a home movie," Bob says. "When I think back to a day in my past, I feel exactly how I felt. I can feel the weather—if it was hot and sticky I'll remember how tight my clothes were and what I was wearing. All of my senses are triggered and I'll remember who I was with, even what I was thinking, my views or attitudes. Sometimes I'll remember something from when I was younger and think, 'Wow I really thought that?'—everything is stirred in my imagination."

As our waitress guides us to our table, Bob tells me about his childhood. The middle of three brothers from western

Pennsylvania, he was a young teenager when he first noticed his memory was different from other people's. "I would talk to my buddies about something that had happened when we were kids," he says. "I'd say, 'Yeah, remember, it was on February fourth, it was a Friday.'"

It became a bit of a party trick. "People often misunderstood it. They called me Rain Man, but it was just a weird quirk to me—you know, like being left-handed or something. I didn't think it was rare, I assumed maybe a few million people had it."

I wanted to test Bob for myself. In 2013, I'd had a quick chat with him over Skype, while researching an article about memory. At the time, I'd asked him what had happened two years earlier, on November 7, 2011.

"Okay," he'd said. "Do you remember what you did that day?"

I'd thought for a second and said no. Despite having picked the date of my birthday, I didn't remember.

"Well, it was a Monday," Bob said. "It was the day after my favorite team, the Pittsburgh Steelers, had lost to the Ravens on Sunday night. I remember waking up on Monday and feeling bummed out about that. I was working up at Cape Cod, in Massachusetts, wrapping up on a show called *Reel Men*. That evening I emailed an ex and she replied the next day."

Back in 2015, sitting at our table in Truxton's I decide to ask Bob about that same day, November 7, 2011.

"That was a Monday," he says immediately. "It was the day after the Steelers lost a close game to the Baltimore Ravens. I was working on a job in Cape Cod. A show called *Reel Men* about fishermen searching for giant tuna. I couldn't get to sleep that night, and sent an email to an ex-girlfriend. I was hoping she would send me a response back, and the next morning she did, and I felt content the rest of the day."

I am startled. What on earth is going on in his brain that isn't happening in mine?

To find out, we have to go back to the 1950s and into a surgical room at the Montreal Neurological Institute and Hospital in Canada. There we find Wilder Penfield, a pioneering brain surgeon wielding electrical currents alongside his scalpel. While performing operations on people with epilepsy, Penfield made use of the fact that their brain was exposed while they were fully awake to find out what happened when he stimulated different areas with a small electrical current. During one operation, he stimulated an area overlying the hippocampus within a young woman's temporal lobe. Suddenly she spoke: "I think I heard a mother calling her little boy somewhere," she said. "It seemed to be something that happened years ago . . . in the neighborhood where I live."

Penfield stimulated the spot again, and once more the mother's voice cried out. He moved the stimulus a little to the left, and the woman heard different voices. It was late at night, she said, they were coming from around a carnival somewhere—some sort of traveling circus. "There are lots of big wagons that they use to haul animals in."[10]

The tiny jolts of activity applied by Penfield seemed to be bringing to life long-forgotten memories—like reaching into a dusty album and picking out a photo at random.

The current theory accepted by most neuroscientists is that memories actually live at synapses—the gaps between neurons where electrical impulses pass from one cell to the next. When these impulses pass repeatedly between two neurons, that particular synapse is strengthened, and any further activity in the first neuron is now more likely to stimulate the second. It's a bit like walking through a dense wood. The more people walk along the same path, the more it becomes clear and the more likely that path will be used again. It works the other way around too, if neuronal pathways aren't being used, they degrade just like real pathways. And that's why we find ourselves forgetting things that we don't practice or think about time and time again.

Much of this activity occurs in the hippocampus, but it doesn't work alone. Think of being handed a bunch of flowers, for example. Henry Molaison proved that forming a short-term memory of this event wouldn't involve the hippocampus at all. In fact, this event would be processed by parts of the cortex that are responsible for touch, vision and smell. It is when these events need to be remembered for more than thirty seconds or so that the hippocampus springs into action and we see the connections between relevant areas of the cortex and the hippocampus strengthen and grow, helping us permanently scribe the memory into the architecture of our brain.

The hippocampus seems to glue different aspects of a memory together. Indeed, when people attempt to learn new associations and recall them later, those whose hippocampus generated the most activity while learning the associations are best at recalling them later. It's as though they managed to stick them together better in the first place.

As such, I tend to imagine our memories as a kind of spidery web of neurons—one that stretches across different areas of the brain and that strengthens and weakens over time. The stronger and more numerous the connections, the more vivid that memory will be and the more easily you will recollect it. Break that web and your memories disappear forever.[11]

Although Bob's memories appear to be more tightly knitted together than mine, just like my memories some days feel more vivid to him than others. For the majority of us, our most vivid memories are those that have some kind of emotional content. When we feel aroused by love, stress or fear—anything mildly stimulating—our brain releases stress hormones that stimulate the amygdalae, two almond-shaped structures involved in our emotional behavior. The amygdalae then send messages to many brain regions to increase the strength of synapses at work at the time. They essentially tell the rest of the brain, "these events are important, remember them." This in turn makes the memory of that event more easily recalled at a later date.

When I think of my most vivid memories, one of the first things that comes to mind is a Bon Jovi concert back in 2013 in Hyde Park. It was midsummer and a beaming hot day. I was with two of my best friends, there was prosecco, sunshine and an electric atmosphere. I remember feeling extraordinarily happy. The next moment that comes to mind is seeing my eldest sister try on her wedding dress in front of my parents in their bedroom. I had to walk out of the room, overcome with emotion. Suddenly I am holding my husband's hand at our own wedding, watching our nephews play football outside a giant teepee while our friends mingle in the sun.

I ask Bob what his most vivid memory is; his answer surprises me. It isn't a wedding, or a birth or a traumatic experience—it is just a nice, normal day. It's May 7, 1970, to be precise.

"This day really sticks out," says Bob. "I remember it very clearly. I was twenty, I was in college full time and I was also working in a mental health center as an orderly. On March thirteenth that year I had done some impressions in class and people loved it, so I was being taken up to the main campus to do a speech course. It was a beautiful spring-weather day. I went to six o'clock mass because I had to work seven till three and I remember walking to church up these steps and being very aware of how happy I was. Then I worked and went to a bowling class.

I went home and drove my car up to the branch campus and we met the professor and two other students. I'd never been up to the main campus and it was hustle and bustle and beautiful and I was very aware of that. I remember the whole day and all my feelings, as well as the cool breeze against my face. It was just a really nice day."

It made me wonder why the rest of us don't remember more of this mundane stuff. Is there any benefit to forgetting?

The American psychologist William James said in the late nineteenth century that if we remembered everything, we should on most occasions be as badly off as if we remembered nothing.

Most of our autobiographical memories, he said, go through a process of foreshortening—we omit facts and emotions associated with our past and generalize things that have happened to us. This explains why I can't remember whether or not I've turned off the gas under the kettle: if you do a task regularly, your memories of that task merge together. Because of this, most of the fine details are lost amid an ocean of generalizations, meaning we find it hard to pick out the more mundane experiences of our past. A little trick I have subsequently learned is to make a different animal noise (out loud) when switching off the gas under my kettle. It feels silly at the time, but it makes the action of turning the gas off much more memorable when trying to recollect whether you've done it later in the day. The animal noise prevents the memory from being grouped within the sea of similar experiences.

You wouldn't want to do this all the time. We use memories of our past experiences to help guide our decisions about the future. If we were able to recollect our past in any great detail, it could take us forever to sort through it all. "Oblivion," said

James, "except in certain cases, is thus no malady of memory, but a condition of its health and its life."[12]

Having learned this, I wasn't surprised to hear that Jill struggles with her daily bombardment of memories. It has led to several bouts of depression. She often feels desperately sad, said McGaugh, constantly remembering the worst times in her life.

Normally, people don't tend to dwell on the past, but Jill's constant recollections seem to link one event to another in an unstoppable manner. McGaugh knows of no other person who is "both warden and prisoner of their memories."

I ask Bob whether he's ever met Jill. "No," he says, "but from what I've heard, her memories seem to consume her life. She's written that she feels haunted by the never-ending stream of memories that appear in her mind. Thankfully, it's not the same for me, or for the other people with HSAM that I've met."

Indeed, most of McGaugh's small tribe don't tend to think of their minds as cluttered—they actually seem to enjoy organizing their memories. They appear to be able to pull out memories at appropriate times, flicking through the past either for enjoyment or out of necessity.

"But what about the painful memories?" I ask Bob. "Isn't it horrid to remember them so vividly?"

"When you remember painful memories and they feel as if they happened yesterday, you can see why it could be awful if that's all you think about. And sometimes when something bad has happened once, if you're in a similar situation, you start reliving that past memory and it makes you anxious that it will repeat itself. But I think one of the benefits of remembering the bad stuff so vividly is that you can learn from your mistakes more easily than other people can."

"In what way?"

"Just that being able to remember all the details and the way you felt when you made a mistake makes you think, 'Okay wow,

I won't do that again,' in a similar situation. And anyway, most of the time, the bad days aren't really that bad, so I don't dwell on any of that stuff—I like being in the present."

AS WE EAT, WE TALK about school and Bob's early life.

"I remember a lot of stuff from when I was young, but not the dates. I remember a few things from when I was really young. My earliest memory is of my mother holding me in her arms—I was drinking milk," he says.

My earliest memory is also of my mum. Except she was hanging me upside down over a sink in our downstairs toilet, trying to clear my airways during a particular nasty bout of whooping cough. I can clearly recall seeing the sink bounce in and out of view just inches from my nose, and the tight, cramped space of the room. I later asked my mum whether she remembered this event. She said that it could have been one of many times during the month I had the illness; she remembered several occasions having to pull thick phlegm out of my throat with her fingers—I was two and a half years old.

"What were you, like two, three?" I ask Bob. I am assuming that that earliest memory refers to drinking milk as a toddler, but the smile on his face makes me pause.

"I think I was drinking from the breast," he says.

"You are kidding me."

He laughs. "You know I always joke about this, but I think it's true," he says. "I remember she had this contented look on her face. I think that memory would have been around nine months or something, I was definitely a baby."

This intrigues me, a memory from nine months old—was that possible, even for a man who never forgets?

Our earliest memories are normally a blurry echo at best. Several theories have been proposed as to why we have this so-called infantile amnesia. Freud, of course, blamed it on adults suppressing sexual fantasies of early childhood that they were later too ashamed to remember—a theory that has since been discredited. The more likely explanation is that the neurons in the brain responsible for forming memories are growing, maturing and being pruned rapidly during the first few years of life. As new neurons are produced, particularly in the hippocampus, the brain must clear out older memories to make room. When Paul Frankland, a scientist at the Hospital for Sick Children in Toronto, accelerated the production of new brain cells in the hippocampus of baby mice, he discovered that they forgot more. When he did the reverse—slowing the growth of neurons using a chemotherapy drug—the pups remembered more than usual.[13] Another theory is that as young infants we lack self-perception and language skills, which are perhaps necessary to embed our memories into contexts that we are later able to look back on in our adult lives.

So does this mean that Bob's nine-month-old memory was fake? I asked Patricia Bauer, professor of psychology at Emory University in Georgia, and an expert in infantile amnesia. She said there is a wide variation in the age of our earliest memories, from late in the first year to as old as nine. So yes, she said, it was possible to have a memory from nine months old, but that in a typical person we would suspect its accuracy. "It would be tough to say that the memory is of a single event rather than a reconstruction based on lots of similar events in which the person took part, not to mention the huge number of images of infant feeding we view over a lifetime."

So maybe Bob's memory was correct, or maybe it was a culmination of several similar moments that happened early in his life. Regardless, it raised another question: Can we ever trust our own memory?

Mitt Romney once recalled a memory to a crowd of Tea Party supporters about his attendance at the Golden Jubilee, an event that celebrated the fiftieth anniversary of the American automobile and attracted 750,000 people. The jubilee was famous for including one of the last public appearances by Henry Ford. The problem was, it took place on June 1, 1946—nine months before Romney was born.

Was he lying? The Republican Party leader said his memory was "hazy" and that he was four or five at the time. In fact, it was likely that he had heard the story from his dad and it had inserted itself within his own memory, forming what he later considered a true recollection of the event.

It wasn't until the 1990s that researchers really started scientifically testing the idea of false memories, when the cognitive psychologist Elizabeth Loftus, then at the University of Washington, wrote about an experiment her team had performed on a boy called Chris.[14] The fourteen-year-old Chris described an experience of visiting a local shopping mall in Washington when he was five. He remembered the visit in incredible detail because he got lost after walking off to visit a toy store. Unable to find his family he thought, "Uh-oh. I'm in trouble now." He remembered thinking he was never going to see them again. Eventually, an old man with a bald head and a flannel shirt helped reunite them all.

The strange thing is, most of this story never actually happened. It was made up by Jim, Chris's elder brother, in collaboration with Loftus. Jim had fed Chris some of the basic facts of the story—the old man, the mall—but Chris had filled in the rest of the details. Chris's story showed that it is possible to plant completely false memories into a person's mind. Since then, Loftus and others have repeated this experiment, implanting all sorts of fake memories, from choking and near-drowning to demonic possession.

Even the most educated minds can be manipulated. When Loftus was just fourteen years old, her mother drowned in a swimming pool. On her forty-fourth birthday, Loftus attended a family gathering at which an uncle informed her that she had been the one to discover her mother's dead body. Although she had previously remembered little about her mother's death, suddenly memories of the incident came flooding back. A few days later, Loftus's brother called her and told her that their uncle had made a mistake—it had actually been an aunt that had found their mother. The memories that had appeared so clear and vivid for the past few days were entirely false. Loftus's own experiment had been inadvertently performed on herself.

Fake memories can have some serious consequences. On November 15, 1989, fifteen-year-old Angela Correa went missing from school. A few days later her body was found, raped and strangled to death. Suspicious of seventeen-year-old Jeffrey Deskovic, a student who had been absent from school during the time Correa went missing, police brought him in for questioning. After being interrogated for six hours, he finally confessed to the murder. Although DNA testing did not match Deskovic to the crime, he was convicted on the basis of his confession and faced life in jail. Sixteen years later, new DNA evidence matched the crime to Steven Cunningham, a man who was serving time for another murder and who subsequently confessed. Deskovic was pardoned and released.

You may find it inconceivable that a false confession could be extracted. But it happens on a surprisingly regular basis. A campaign group in the United States called the Innocence Project proposes that false confessions play a role in almost a quarter of U.S. convictions. Perhaps you think you're impervious to this kind of manipulation, but you'd be surprised just how easily you could succumb. Recently, Loftus showed that a lack of sleep can cause people to make false confessions after getting students to admit to pressing the wrong button on a computer task, wiping out a week's worth of data. In fact they hadn't done any such thing, but half of the students who hadn't slept the night before the task believed they had a memory of the event, and signed a confession—compared with less than a fifth of those who'd had a good night's sleep. Tiredness, low IQ, leading questions, all of these things can help convince us to forge a memory of something that never happened.

These examples reveal something quite extraordinary: Our memories, once formed, are not fixed. Each time we retrieve a memory, we strengthen the neural pathways that have created it, and in doing so, reinforce and consolidate that memory so that it becomes lodged more permanently in our minds. But for a short time during this retrieval process, our memory becomes malleable—we are able to reshape it and sometimes contaminate it.

I wondered, Was this the secret behind Bob's incredible memory? Was there something special about the way he retrieved memories that allowed him to strengthen and consolidate them more accurately, and more permanently, than the rest of us?

"It was Billy Mayer," Bob was saying. "They thought he was involved with a girl called Katrina Young. His wife had left him at the time and they became friends and that's where the scandal came from. But there's never been any proof that it was true. People looked into it, but no one proved they were seeing each other. He wasn't seeing her, but it was pretty bad for the town—"

I must have looked puzzled because he stops mid-sentence.

"I'm sorry, sometimes I have to think about what I'm saying." He laughs.

It turns out that Bob is talking about Holland College, a whole community based around a school basketball team called the Golden Knights—a powerhouse in collegiate sports that has played in several championships and boasts numerous high-profile sportsmen such as Otis Pooky and Isaac Moseley. Bob is the Knights' biggest fan, because the team—in fact the whole community—exists only in his imagination.[15]

It started when he was young. Bob decided to create his own imaginary basketball team. Each of the players lived in a place called Tiger Town, and he would play whole basketball games in his mind. The team would compete in championships, and they would win and lose. He thought he would stop, but as he aged, so too did the team. The players worked their way through college, then went off and married and had kids. Nowadays, most have full-time jobs, some have died in tragic accidents, others from old age. "It's like a fifty-year-long book in my brain," says Bob.

If this sounds obsessive to you, it's because it is. Bob has many obsessions—he's a self-confessed germophobe, too. "If I drop my keys on the ground I'll scrub them under hot water," he admits.

These obsessions were the vital clue that McGaugh had been searching for. It soon became apparent that other people with highly superior autobiographical memory (or HSAMers as they like to call themselves) had some kind of obsessive compulsive tendencies. For Jill it was her diaries—sometimes her writing was so small and packed so tightly on the page that she couldn't read it back. For others it was remembering where and when they had first worn a pair of shoes, or cleaning, or watching certain television programs over and over again. Most of them also appeared to enjoy organizing and replaying their memories in some way or another. For instance, when Bob is stuck in traffic, he tries to recall his favorite memories of that particular date—say, every March 1 since he was five. Or he might try to remember what happened every day in June 1969.

"This obsessiveness was a really intriguing part of the puzzle," says McGaugh.

To find out more, he got some of his growing tribe of HSAMers—who are now more than fifty strong—to take part in various tests that stretched other aspects of their minds, such as verbal fluency and ability to memorize faces and names. He wanted to see if they excelled at anything else.

Unfortunately, the results were inconclusive. Just like Jill, the HSAMers weren't much better than others their age at any of the tests—they certainly didn't excel in any. So McGaugh tried another tack. He asked his participants to recall events that happened on each day of the preceding week, as well as a week that occurred one month, one year and ten years earlier. A month later, McGaugh surprised his participants by asking them to recall the same dates, allowing his team to check the consistency of their memories.

As you might expect, those with HSAM had a far superior memory of days further in the past. What was surprising, though, was that both groups could remember the same quality and quantity of information from the preceding week.[16]

That was enough to convince McGaugh that Bob and the others are no better than you or me at acquiring memories; they are not superior learners—they are simply better at retaining them.

McGAUGH WANTED MORE CLUES to the puzzle. So next he scanned their brains. There he found some subtle differences in the structure of nine regions, including an enlarged caudate nucleus and putamen. This was particularly intriguing, because both of these areas have also been implicated in obsessive compulsive disorder (OCD).

Was it a coincidence? The universe is rarely so lazy, said Sherlock Holmes. McGaugh too reckoned there had to be more to it.

It seems, he says, that the initial process of translating an event into synaptic activity, what's called "encoding" a memory, in HSAMers is no different from how it works for the rest of us. So too is the mechanism they use to retrieve a memory. The difference between people with HSAM and the rest of us seems to be happening between the encoding and the retrieval process—a point we call consolidation. Perhaps, says McGaugh, their extraordinary powers of memory are rooted in an unconscious rehearsal of their past. Not that Jill and Bob and the others actively try to memorize their past in order to remember it—that would take some considerable dedication. Instead, he believes that they accidentally strengthen their memories by habitually recalling and reflecting upon them.

"It could be a unique form of OCD," he says.

At the time of this writing, McGaugh is eighty-five, and after more than fifty years of memory research is near to retirement. He is a man who is clearly passionate about finding out what gives a handful of people such an incredible memory. I was interested to know why he had dedicated so much time to such a niche talent.

"These are not small effects," he says. "Their brain must be working in a different way from those of others." He wonders whether it was an ability that all humans had early on and then had lost because there was no pressure to retain these memories or whether it was perhaps an abnormal genetic condition that sprang up from nowhere. "Whatever it is, it is quite remarkable," he says. "How the hell does it happen? That's the question. And that has always been my quest—to understand this marvelous machinery that we call the brain."

Something Bob said to me as we finished our meal stuck with me. "You know, one of the best things about having a perfect memory," he said, "is the ability to remember those I have lost.

"I make sure I think a lot about people I love when they're alive so that I can go back to any time in their life that I was with them and remember it like it was yesterday. Then if they're no longer with me, it's like I can still spend time with them. The people I've lost don't feel like they're truly gone because my memories of them are so clear. I can go back to my younger brother's life and not have to mourn him like others do, because I can remember our times together so well. I think about people a lot and appreciate my time with them because once they're gone, they won't be here, but my memories always will be."

I've thought about that a lot since we met. When my mum was diagnosed with incurable breast cancer, it motivated me to spend the last year of her life making a special effort to concentrate on the time we were spending together. Those memories, I hope, are there for good.

I know my memory will never be as perfect as Bob's or Jill's. But others, like Mullen and Koltanowski have shown me that even my ordinary brain is capable of remembering so much more than I ever realized—just by creating a special palace around my memories that can never be erased.

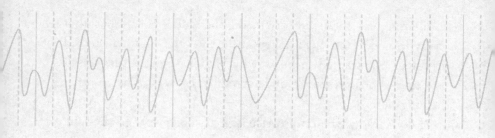

SHARON

Being Permanently Lost

1952

Out on the front lawn, Sharon was blindfolded while her friends ran around her, laughing, trying not to be caught in a game of blindman's bluff. Sharon grabbed hold of someone's sleeve and whipped off the scarf that covered her eyes. "You're it!" she shouted.

Then she blinked and looked around her. Suddenly, panic started to rise. The house, the street, it all looked different. She had no idea where she was.

Sharon ran into the back garden and discovered her mother sitting in a lawn chair.

"What are you doing here?" Sharon asked. "Whose backyard is this? Where am I?"

Her mother looked at her, puzzled.

"What's wrong with you?" she asked her daughter. "This is our house!"

Sharon was totally disoriented. She told her mother that everything around her looked different. Her mother looked irritated. She wanted to know why Sharon thought this wasn't her home. Sharon didn't understand: Why wasn't her mother helping her?

"I don't know where this place is, it all looks wrong," she said. "I'm so confused."

Her mum looked her in the eye, and pointed a finger at her face.

"Don't ever tell anybody about this," she said. "Because they'll say you're a witch and burn you."

PRESENT DAY

"I can remember that moment as if it were yesterday," Sharon says, over the phone. "I was five years old."

Sharon had woken up the following morning knowing that something weird had happened again. She says it was as though her walls had moved in her sleep. She was in her bedroom but things didn't look like they were in the right place. Her door was on the wrong side of her for a start. "I knew it had to be my bedroom," she says, "and bits of the room were familiar, but it was all wrong at the same time, nothing was where I thought it should be."

Sharon didn't know it at the time but her brain was having trouble producing an accurate mental map of her surroundings.

Sharon's disorientation began to occur more and more frequently until it became a constant presence throughout her day. It made finding her way around her neighborhood and her school completely impossible. In spite of this, she never mentioned her problem to anyone. Instead, she used a natural sense of humor and sharp intelligence to complete her education, make friends and even get married without anyone ever knowing she was almost permanently lost.

"I hid it for twenty-five years," she says.

"Twenty-five years?"

"Yes, well, you know . . . the witch thing."

Sharon's condition was one of the most peculiar I had ever come across. The loss of an ability that I had never given a second's thought to—the ability to orient.

I had first heard about the condition after it was described in a medical journal called *Neuropsychologia*.[1] One of the authors of the paper had been kind enough to put me in touch with Sharon, whose case was among the most severe he had discovered.

Eager to find out more about this mysterious disorder, and to include Sharon in my journey, I emailed her. Would she mind if I came over to Denver to meet her in person?

"That would be wonderful!" she replied.

I was keen to see her in her own home. Even there, she said, she can get lost walking between her bathroom and her kitchen.

Just hours after saying farewell to Bob, I grab some sleep in a musty motel that smells of damp clothes and cheese and wake at the crack of dawn to get back to the airport. Bleary eyed, I arrive in Denver. My phone *beeps* as I sit in the parking lot familiarizing myself with the left-hand drive of my rental car. It's a text from Sharon: "I hope you can find your way here without a problem. Call me if you get lost . . . maybe I'll be able to guide you. Ha, what am I thinking!"

I smile and switch on the GPS. The screen flashes and then goes dead. Eventually I manage to get the map to appear— although the image is dark and fuzzy. The irony isn't lost on me, despite my jet lag.

Several wrong turns later, I pull into a quiet neighborhood full of neat little condos. I work my way around the maze of streets and spot Sharon, standing on her veranda, waving down at me.

I announce my arrival to her neighbors by starting the car alarm while attempting to turn off the engine and change out of my driving shoes. I am wearing only one sandal when Sharon opens the car door. It wasn't exactly the first impression I had intended to make. Nevertheless, she greets me with a warm hug and a huge smile on her face.

"It's so great to meet you finally. Aren't you just adorable!"

Sharon has flaming-copper hair, swept into a stylish crop that is set off by a bright-pink blouse. The colors complement her deep-red lipstick. Her sunglasses instantly remind me of one of those slightly eccentric grandmas you see in Hollywood movies.

I surreptitiously slip on my second sandal and follow her to the front door, where we are met by a giant metal lobster with the word "Welcome" written across his rusty belly.

Sharon shows me into her house, which is open plan, peaceful and as neat as a pin. She offers me a drink, and we wander into the kitchen, where I stop to stare at her fridge. There's the normal assortment of memorabilia stuck on the door—pictures of friends, magnets, telephone numbers, notes from grandchildren, a picture of Wonder Woman—but it's a large piece of paper right in the center that catches my eye.

It is a photo of a handsome young Italian, with thick eyebrows and three-day-old stubble, looking into the distance. It is held up with a magnet that says: "A true friend knows everything about you . . . and likes you anyway." A smaller photo, of Sharon and the same man sitting at a dinner table together, arms around each other's shoulders and smiling at the camera, is pinned above it.

"Who's that?" I ask.

"That's Giuseppe. Isn't he cute? He's such a gentle and com-passionate man. He changed my life."

AS A YOUNG POST-DOC, Giuseppe Iaria was fascinated by nav-igation. His interest had begun as an undergraduate student, when he worked on a project investigating why people with damage to one side of their brain sometimes have problems navigating. Later, while working at the University of British Columbia, he decided to investigate why some healthy people have a better sense of direction than others. One day, completely out of the blue, a middle-aged woman, whom I will call Claire, showed up at his lab complaining of a peculiar problem: she was constantly lost.

Iaria suspected that Claire's disorientation was the result of some other condition. He began ruling out possible options one by one. He knew, for instance, that inner-ear infections can damage a delicate tissue called the labyrinth, causing the sen-sation that your world is moving around you. Perhaps it was this, he thought, that was causing Claire to feel disoriented? Brain tumors, lesions and dementia can damage the hippocam-pus, which, as we know, is involved in many types of memory. Could one of these things be preventing Claire from being able to remember her way around? Or maybe it was epilepsy that was stopping her from being able to memorize directions; sud-den bursts of uncontrolled electrical activity in the brain can do that. It took Iaria and his mentor, Jason Barton, two years to cross off all potential problems. But, as far as their tests showed, Claire was in perfect health.

Claire told Iaria that she hadn't lost the ability to orient her-self; she'd just never learned it in the first place. She recalled that from the age of six she would panic at the supermarket each time her mother disappeared from view. During school, she would have to travel with her sisters or parents, and she never left home by herself because she got lost each time she tried. As an adult, Claire had figured out how to get to work by taking a particular bus, and memorizing the stop and a prominent landmark near her office. But her work was moving to an unfamiliar area, and she had decided it was time to get some professional help.

Iaria was intrigued. He routinely encountered disorientation as a symptom of other conditions, but never as a developmental disorder—one that occurs as you grow up. Determined to get to the bottom of the problem, he took Claire out for a short walk around the local area. He then handed her detailed directions for how to repeat the route by herself. Claire followed the directions without any mistakes. However, when Iaria asked her to draw a map of the route she had just walked, or of the town in which she lived, she found it impossible. She said she did not have "in my mind a map to report."[2]

Iaria called her Patient One and named the condition developmental topographical disorientation disorder: the inability to generate, and therefore use, a mental map of your surroundings, despite an absence of any brain damage.

Iaria figured there must be others out there with the same condition and created a website to encourage people to test their navigational skills. He also went on a radio chat show to talk about the disorder. In the middle of the broadcast, he received a phone call live on air.

"It was like we'd staged it," he told me. "A guy called in and said, 'I'm always lost. I've always been this way. I've told people and they just don't get it, they think I get distracted and I give up. I don't tell people anymore, they just don't believe that I can be that bad at directions.'"

Over time, Iaria found others. One person told him: "No

matter how long I live in the same building I can never picture in my mind where the washroom is."

Sharon was case number four. Unfortunately, by this time, she was sixty-one years old.

I settle on the sofa with a glass of water. Sharon sits opposite.

"Take me back to the beginning. Were you permanently lost from the age of five?"

"No," she says. "Some of the time my world looked perfectly normal and I could navigate perfectly well. But then all of a sudden my world would flip, and I'd become completely disoriented."

"And you never told anyone?"

"No. Instead I was the class clown. I thought if I could stand up and make the class laugh, they wouldn't know my secret, so I became this comedienne."

"So no one ever noticed that most of the time you were completely lost?"

"No. I would just follow my friends when we walked to school, and if it happened during class, I'd spend the rest of the lesson trying to memorize the way the room looked so that I'd know where everything was the next time it happened."

One day, when Sharon was still a young girl, she came across a solution. She was at a friend's party, and next in line to play Pin the Tail on the Donkey.

"You know the one," she says. "Where you're blindfolded and you spin around and then try to stick the tail in the right place. After I spun around I just knew something was horribly wrong. I felt like I was walking in completely the wrong direction. I pinned the tail on the donkey and everyone laughed like they do, and I took the blindfold off. I thought, 'I know I'm at my friend's house, but this doesn't look like my friend's house.'"

This momentary crisis turned out to be the saving grace that would help her navigate the rest of her life. For when it was her turn to be blindfolded and she was spun around for a second time, Sharon's world flipped back to normal.

"That's when I learned that spinning could cause the disorientation to happen. But that it also fixed it."

"These days I usually try to find the nearest bathroom," Sharon says. "I go into a cubicle, close my eyes and spin around. I can't quite describe the sensation. It's not a sound, just a sense that everything feels back to normal. When I open my eyes my world is recognizable again."

She chuckles and points toward the picture pinned to her fridge. "I call it my Wonder Woman impression."

"Why do you do it in the bathroom?"

"Well, what would you think if you saw an old woman standing by her car spinning around in circles with her eyes closed?"

She has a point.

"I always did it in secret because I was humiliated by it all."

For most of us, navigating feels easy and automatic. You arrive in a new city and your brain starts trying to make sense of the place. On your first day, you find your home, your base for the trip, and over time you start recognizing certain landmarks. You become familiar with your surroundings.

Many of Iaria's patients feel like they live in a constant "first day." No matter how much time they spend somewhere, their surroundings never become familiar.

Like Claire, many have learned to navigate the most important routes in their life by remembering a specific sequence of turns. To get from their desk to the bathroom, for instance, they know they need to turn left at the printer, right at the potted plant and go through the double doors.

But there's a reason this isn't how you or I navigate. To remember all your journeys this way would place a huge strain on your memory. Instead, we use a dynamic tool, what scientists call a cognitive map, a kind of internal representation of our surroundings that becomes familiar so that we don't have to remember a specific sequence of directions, but can merely visualize where things are in relation to one another and ourselves.

Try it for yourself now. If I asked you to think about the route you'd need to take to get to the bathroom, could you do it? You probably don't even have to try. We tend to take for granted the ability to picture a route in our mind's eye, but it is a remarkable skill—in fact, one of the most complex behaviors that our brain can perform. And one that has puzzled scientists for decades.

Part of the problem is that normal navigation enlists several areas of the brain, all of which are having an incredibly sophisticated conversation with one another.

WHEN SHE'S NOT SCANNING world memory champions, Eleanor Maguire spends a great deal of her time trying to work out exactly what parts of the brain are doing the talking. Her motivations aren't completely selfless—although she is one of the UK's top navigation researchers, she is also hopeless with directions.

"It was absolutely the reason I got into this area of research," she said, when I dropped by her lab one day. "I'm so bad at directions—it's really debilitating."

We were sitting in her office, in Bloomsbury, central London. Maguire tells me that when she goes out the front door she will deliberately turn in the opposite direction to where she thinks she needs to go. That way, she said, "I'm correct about ninety percent of the time."

One afternoon not long ago, I hurried past Maguire's lab on my way to get my hair cut. I was late so I rushed to the main road and stuck out my hand. I was rewarded by the appearance of Geoff, a black-cab driver who had been ferrying people around the streets of London for more than twenty years. I climbed into the back seat and reached for my seat belt.

"Where to, love?" he asked.

"South Molton Street," I replied.

Without a second's thought, Geoff spun the cab around, slipped down a side street and headed directly to the salon. He never once consulted a map. That's because he had mastered the Knowledge, a famous test that all London cabbies must pass, which involves learning all twenty-five thousand roads within a six-mile radius of Charing Cross station.

Maguire wondered whether cab drivers like Geoff, who have an incredible sense of direction, might reveal what makes other people so good at navigating. By scanning their brains, she discovered that the back part (posterior) of the hippocampus in taxi drivers is larger compared to that of non-taxi drivers.[3] But was that a result of being a taxi driver, or were people with bigger hippocampi just more likely to become taxi drivers? To find out, Maguire scanned the brains of seventy-nine trainee taxi drivers several times over four years as they began to learn the Knowledge. Those who passed the test had a bigger posterior hippocampus than when they started, whereas there were no changes seen in trainee taxi drivers who failed their exams or in thirty-one people whose age, education and intelligence were similar to the taxi drivers', but who had never attempted to learn the Knowledge.[4] Clearly, the hippocampi were growing alongside navigational abilities, which raises the following question: How do they help us get from A to B?

IN THE 1960S, the British neuroscientist John O'Keefe, also at University College London, began dabbling with the idea that the secret to normal navigation lay in the hippocampus. To test this theory, O'Keefe studied the brains of rats as they walked around an open space. He wanted to know which neurons were active as the rodents explored their environment. He did this by placing a set of thin electrodes into their hippocampi, which could record the little spike of electricity that occurs when an individual neuron is communicating with its neighbors.

Using this technique, O'Keefe discovered a type of cell that fired only when the animal was in a specific location. Each time the rat passed through this location—pop!—that cell would fire. A nearby cell seemed to care only about a different location. Pop! It would fire whenever the rat walked through that location. The next cell would respond only to another location, and so on. Pop, pop, pop! The combination of activity of many of these cells could tell you exactly where that rat was to within 5cm^2. O'Keefe named them place cells and showed how together they told the rest of the brain, "This is where I am in my environment right now."[5]

Over the next few decades, it was discovered that place cells don't do this job alone. They receive input from three other kinds of cells in a nearby region called the entorhinal cortex. One type of cell is called a grid cell, and was discovered by May-Britt Moser and Edvard Moser, former husband-and-wife team, both born on remote islands on the outer west coast of Norway.

The Mosers realized that our ability to navigate partly relies on us being able to think about how we are moving and where we have come from. Consider the way you head to the ticket machine in a parking lot, and then reverse the movements of your body to return to your car. The Mosers discovered that grid cells were the neurons responsible for integrating this information into our cognitive map.[6]

To understand how grid cells work, imagine running around a carpet that is covered in a grid of interlocking hexagons, a bit like a beehive. One grid cell will fire whenever you reach the corners of any hexagon in that grid. Shift the grid along ever so slightly to another section of the carpet, and another grid cell will be responsible for firing every time you reach the corners of that grid's hexagons—and so on. These cells establish a generic map of space, giving you constantly updated information about your position and the relative distance between certain landmarks.

The entorhinal cortex is also home to border cells. These cells give you information about where you are in relation to certain walls and boundaries. One might only fire, say, when there's a wall that is close by to the south. Another might fire when you're midway between two walls, or when you're near the edge of a cliff.

To complete the picture, border cells also share their real estate with head direction cells, which, as the name implies, fire only when an animal's face is turned in a specific direction.

The most widely accepted theory of how we find our way around the world is that our brain stores patterns of how place cells fire in a specific location, so that they can act as a guide for when we return to that location. Imagine looking for your car after a long day's shopping, for instance. Your place cells crackle away, influenced by the direction of your head, the movement of your body and the landscape around you. They guide you around until the current pattern of activity matches that of the stored pattern, and voilà!—you've found your car.

That's not the end of the story, however. Despite all this activity, our internal compass is incomplete. We're still missing a piece of the navigational puzzle—one so important that its loss can mean the difference between life and death.

When you find my body, please call my husband George and my daughter Kerry. It will be the greatest kindness for them to know that I am dead and where you found me—no matter how many years from now.

WHEN SIXTY-SIX-YEAR-OLD Geraldine Largay stepped a short way off the Appalachian Trail to relieve herself, she never expected that she wouldn't be able to find her way back. Known to her friends as Gerry, she was a retired air force nurse who had hiked other long trails near her home town in Tennessee. She had taken a course on how to traverse the entire Appalachian Trail, which runs over 2,200 miles and crosses fourteen states, and had completed more than 1,000 miles of the six-month journey.

On July 22, 2013, Gerry tried to text her husband, who was waiting at a nearby checkpoint, ready to deliver her fresh supplies for the next leg of her trek. "In somm trouble," it said. "Got off trail to go to br. Now lost. Can u call AMC to c if a trail maintainer can help me. Somewhere north of woods road. XOX."

Lack of signal meant the text never sent, so Gerry camped down for the night. The following day an official search began. For weeks they scoured the thick woodland to find her.

In October 2015, a forester working for the U.S. Navy discovered a human skull with a sleeping bag around it. According to a report in the *New York Times*,[7] a flattened tent lay a short distance away, along with a green rucksack containing Gerry's possessions, neatly packed into ziplock bags. Nearby was a moss-covered notebook that said "George Please Read XOXO." In the notebook Gerry explained that she had walked for two days after being unable to find her way back to the path. As she had been taught in her training, she set up camp in the hope that someone would find her. The last entry was dated August 18, 2013.

While it is impossible to say what Gerry could have done to avoid this tragic fate, her disorientation was no doubt aggravated by the fact that she left the path at one of its most rugged sections. She wouldn't have had to walk far before she was surrounded by thick underbrush and identical fir trees packed so tightly it would have quickly become impossible to make out the trail. There was nothing that distinguished one direction from the other. In short, there were no landmarks.

You might not think much of being able to remember that there's a post box at the end of your street, or a bus stop outside your office, but in fact the ability to recognize and incorporate these permanent landmarks into your internal map of the world is incredibly important. We constantly fill our mental maps with things that are meaningful to us. Think about directing someone from the nearest station to your home. What features of the route would you use to keep them on track? I'd use my local art deco pub, a museum containing an overstuffed walrus, and a distinctive triangular hill under which victims of the plague are buried.

Our ability to recognize familiar landmarks is so important that there's a part of the brain that is dedicated to the task—it's called the retrosplenial cortex and when it's damaged, it leads to severe problems in navigating.

WHEN WORKING CORRECTLY, our spatial memory is outstanding. But with the use of technology—GPS, satnav, mobile maps in the palm of our hand—could we all lose our ability to navigate? Calculators, after all, have caused many people to suffer a drop in their mental arithmetic skills. In a commentary in the journal *Nature*, Roger McKinlay, former president of the Royal Institute of Navigation, says that this might indeed be the case.

"If we do not cherish them, our natural navigation abilities will deteriorate as we rely ever more on smart devices,"[8] he said.

Our natural navigational skills can indeed be blocked by technology. Studies have shown that people who follow GPS systems to get from one place to another find it more difficult to work out where they have been than people who use traditional paper maps. Like many of our brain's talents, it's a case of use it or lose it. In 2009, Maguire and her colleagues showed that London taxi drivers who had recently retired performed worse on navigation tests than drivers the same age who were still guiding passengers around the capital.[9]

Whether our technological crutches will eventually erode our natural navigational skills is unclear. A much more pertinent problem comes from not noticing that your tech is taking you somewhere you don't want to go. In 2013, an elderly Belgian woman set out on the thirty-eight-mile journey to her home in Brussels, but after mistakenly entering the wrong address in her GPS arrived in Zagreb, 901 miles and two days later. Some stories are more tragic. In 2015, on her way to the beach, a businesswoman was shot dead in Brazil after a navigational app made her take a road through a gang-controlled favela. Even the most technologically advanced navigational systems may know where they are, but they don't always know the best way to go.

So are we becoming dangerously deskilled? Putting your faith in your GPS rarely leads to such awful situations, and is unlikely to diminish your natural abilities completely, but it's important to remember that you carry around with you an internal map that is—for now—more powerful than the smartest GPS.

Sharon and I set off for lunch at a nearby restaurant. I offer to drive, but Sharon insists she knows the way and that it won't be a problem. These were lines delivered with confidence. But should a woman who has difficulty finding her own kitchen really get behind the wheel?

I'd been keeping a watchful eye on Sharon as she showed me around her house. I don't know what I was expecting—perhaps for her to look confused suddenly and then bump into a wall or something. But nothing out of the ordinary had happened, so I happily jump into the passenger seat.

From her condo we drive around a couple of roundabouts, pass through a set of traffic lights and indicate left, then right without a hitch. We turn safely on to a small highway that runs through the town, the snow-topped foothills of the Rocky Mountains dominating the landscape to the west.

Sharon points to the mountains and tells me that sometimes she'll be driving into town when she suddenly realizes that the mountains are to the north and she'll know that her world has flipped. Before I am able to process this statement she points to our restaurant. Then we fly past the entrance. "I can't go in that way because it's a big curved road," she says, as if that's a perfectly obvious explanation for our detour.

As we park, I glance up at the mountains, so solid and unmoving. How could they suddenly face north?

We sit down in Salsa Brava and order two iced teas. I ask Sharon to go back to basics.

"Can you explain exactly what you see when your world flips?"

She pauses for a moment, and then tells me to think about a busy shopping street in London. I choose Oxford Circus, with its heaving crowds and constant flow of human traffic.

"Imagine you've had a busy day shopping," she says. "You come out of the last shop and head left toward the station."

I picture the scene.

"All of a sudden, though, you realize the station is actually on your right because you were in a shop on the opposite side of the street to where you thought you were. In that split second, you feel momentarily disoriented because the station that you thought should be east is now west. Your world hasn't literally flipped but your perception of it has."

When that happens, most people's brains are surprisingly forgiving. As soon as the brain gets confused, it spins everything around and it reorients itself—and you—within milliseconds. But that split second, in which your mental map doesn't match with where things actually are, is how Sharon says she feels when her world is flipped. So when she says the mountains are suddenly to the north, it's because her mental map has shifted them to the north, even though they haven't physically moved an inch.

"I just can't flip my world back around like you can," says Sharon. "Unless I do my Wonder Woman impression."

I ask why we had to take a detour into the restaurant. Sharon explains that curved roads make her world flip. It has made finding work difficult. In her mid-twenties she struggled to find a job. Every time she had an interview, she would have to work out in advance where the building was and whether it sat on a curved road. If there were a lot of winding passages in the building itself, she would have to turn down the job altogether.

I want to know more about what Sharon's alternate world looks like—was it not possible for her to recognize enough of her environment to work out which way to turn?

"It's hard to explain," she says. "Think about standing in front of a bathroom cabinet with a mirrored door. Open that door and look at the rest of the room through it and you'll know it's your bathroom but everything's kind of in the wrong place. Plus you're stressed because everything looks different. It makes it hard."

When Sharon has to get up in the middle of the night to go to the toilet or if she's in a rush in the morning and doesn't have time to do her Wonder Woman impression, she says she feels like she is in a whole other condo. When she had young kids, she would have to follow their cries to find their room if they woke her suddenly in the night.

"When it happens at home, I'll know I'm in my kitchen but I can't tell you what's contained in any of the cabinets or drawers because I don't have any attachment to them. I have to say to myself, 'Okay, pretend you're in your correct kitchen,' and I know when I need a spoon I go to the drawer to the right of the fridge. So I look at the fridge in this 'other' kitchen and I say to myself, 'Okay, that's where the spoons are.'"

THROUGHOUT SCHOOL, SHARON CONTINUED to hide her problem from her friends and family. The seeds of condemnation planted by her mother at such a young age had clearly taken hold. I feel a wave of sympathy. Sharon is so likable—so friendly, funny and intelligent. I'm amazed she kept this to herself for so long.

Sharon was almost thirty before her secret came out in the open. Her brother had phoned her asking to be taken to the hospital. He suffered from Crohn's disease and was feeling unwell. Sharon rushed out of the house in a panic, got into the car and set off on the short journey to his house. But somewhere along the way, her world flipped and she got completely lost. She pulled into a gas station to use the phone.

She called her brother. "I can't find your house," she said, and described the gas station.

Her brother was confused. He said, "You're two blocks away from me—how do you not know where you are?" After the two of them had returned from the hospital, her brother asked her what was going on.

"It was so emotional for me, I could hardly say the words."

It was the first time Sharon had talked about her condition since she was five.

"After I told him what my mother had said, he was so mad, but he understood why I hadn't said anything. Our parents weren't well—we didn't have a normal childhood."

Sharon's brother told his doctor about her condition, who set up a meeting with a neurologist. With appointments to go to, Sharon was forced to tell her now ex-husband everything. Up until that moment she had managed to hide it from him very successfully.

"I did very little driving, and the places I drove to were only very close to my house," she explains. "I had already routed them out on straight streets so I wouldn't get lost."

My thoughts about her bumping into things weren't completely off the mark. She told me that she had been terrified of not being able to take care of her children in an emergency. When she jumped out of bed to get to her children in the dark, she would almost always run into a wall. Her husband just thought she was clumsy.

"I allowed him to think that rather than trying to explain. I just felt so stupid."

When Sharon finally admitted the truth after eight years of marriage, the only thing she remembers him saying was "Is that why you're always asking me which direction we're driving?"

"He just didn't seem interested."

SHARON'S NEUROLOGIST TOLD HER that since she'd had the problem for so many years, it sounded like she might have a benign tumor or epilepsy. In either case, he said, "We'll put you in the hospital, do lots of tests and try to fix it."

True to his word, he organized a barrage of tests, looking for signs of strange brain activity that would suggest epilepsy or an anatomical lesion that might explain the loss of direction.

"I just thought, please, God, let there be something there that they can fix," says Sharon.

But there was no epilepsy, no lesions. Sharon's brain looked completely healthy.

"They said I needed to see a psychiatrist—they thought I was crazy." The diagnosis caused her to suffer a bout of severe depression.

"I wanted to die," she says. "I'd just had my hopes raised thinking that the doctors would find something that could be fixed."

Sharon saw a psychologist for more than a year, and although he helped her work through her depression, he was unable to fix her disorientation. He told her to keep checking in with a neurologist every few years to see if the research community had discovered anything new. He said, "I truly believe that this is something happening in your brain that we just haven't yet learned about."

It wasn't until Sharon turned forty that she felt strong enough to follow his advice. She made an appointment to see a doctor at the hospital she was working in as an administrative assistant.

But as soon as she sat down, she felt uncomfortable.

"This doctor gets out her little pad and paper and asks me what was going on," says Sharon. "I tried to tell her as simply as I could—that my world lifts up, turns around and sets back down and I'm totally lost. She looked at me like I'm telling her a made-up story. She asked me how I correct it and I told her I spin around and it fixes it. She said, 'Let me see you do it.'"

Sharon was taken by surprise. She had never spun in front of anyone before.

She winces at the memory.

"I swallowed my pride, stood up and closed my eyes. It was so embarrassing. I spun around in circles until I knew the world had flipped."

The doctor asked Sharon what she saw.

"I said, 'Well, I'm in a different room now. I know logically

I'm not, but this doesn't look like the same room as when I came in.'"

Sharon spun around again and sat back down. The doctor put down her pen and pad and said, "Has anyone ever suggested the possibility that you have a multiple personality disorder?"

Sharon was mortified.

"I'd shared my story and felt like I was being told I was crazy again. I just couldn't go through that again. I got my purse and left."

IT WAS ANOTHER DECADE before Sharon made any further attempt to understand what was wrong with her brain. A friend had read some books by the neurologist Oliver Sacks and recommended that Sharon write to him about her symptoms. He replied a few weeks later. Sacks began by apologizing that he had not heard of such a condition. But he said he was reminded of stories told to him by astronauts who said they had experienced times in space when everything "looked wrong," upside down or at an angle. Their world would suddenly right itself, when some clue—often a tactile one—allowed their orientation to become reorganized. Sacks also told her that the problem of not recognizing familiar environments might be similar to another condition called prosopagnosia, in which people are unable to recognize familiar faces.

On hearing this, Sharon took to the internet and googled "prosopagnosia." She came across a website that tested how good you are at recognising faces. After the test, there was a questionnaire. Halfway through, one of the questions hit a nerve: "Have you ever been in an environment that you know you should recognise but doesn't look familiar?"

"I was like 'Holy shit!'" says Sharon, as a bemused waiter places our lunch on the table. "I wrote everything about my condition in the notes section, trying to be as succinct as possible."

Sharon pauses and turns to the waiter.

"She's writing a book about crazy people," she tells him. "And I'm one of them!" She laughs.

Without any further explanation, she returns to her story.

"Within a week, I received a call from Brad Duchaine, a researcher at University College London." Duchaine had created the online test that Sharon had filled in, as part of a project that attemped to work out what brain mechanisms allow us to recognize our friends and family.

"He was so sweet," says Sharon. "He believed everything I said, and assured me that at some point there would be someone doing research on my problem."

"I promise you," Duchaine told her. "When I find out who it is and where they live, I will let you know."

"He just really made me come out of my funk," says Sharon. "He gave me hope that what I had was a real thing, that I wasn't just crazy, that I wasn't a witch."

Later that year, Duchaine emailed her saying he had good news. There was an Italian researcher moving to Vancouver to start researching the condition that she had described. That man was Giuseppe Iaria, and shortly afterward Iaria contacted Sharon to invite her over to his lab.

"The first time Giuseppe called, I sat at my kitchen table and I told him everything. He was such a gentle man, he nearly cried when I told him about the witch thing."

IARIA TOLD SHARON that he thought there might be a problem with the way in which the different navigational cells in her brain communicate with one another. Over the next five years, he began to test this theory.

He started by scanning the brains of healthy people, looking at how different brain regions known to be important for orientation and navigation communicate with one another, and how that communication related to orientation skills. His team concluded that the best navigators were those with higher levels of communication among all of the regions of the brain involved in navigation.

This concept is called network theory and it's an idea that underlies many human behaviors—that the connections between different regions of the brain that allow them to talk to one another may be more important than how well the regions function by themselves. It's like having a quartet of the world's best brass players who individually make wonderful sounds, but if they're not playing in time with one another, that music turns into mayhem.

Iaria's team then went on to scan the brains of a group of people with Sharon's disorder. They noted a difference in the activity of their right hippocampus and parts of their frontal cortex, an area that allows us to draw all the information about navigation together and make judgments upon it. It's also an area involved in reasoning and general intelligence.

Since Iaria's patients had no problem with their memory, or their reasoning, he concluded that the condition must be a result of ineffective communication between the two regions, rather than a defect in each individual region alone.

"It's not enough for the individual parts of the brain to be able to speak," he told me. "They have to have good conversational abilities too."

Since then, Iaria's team has discovered that, just like Claire's, Sharon's brain looks anatomically normal, but several of the areas involved in navigation don't communicate properly. I understood how this could prevent Sharon from forming a mental map of her surroundings, but I was confused as to why she could sometimes navigate perfectly well. "What causes the sudden flipping?" I asked Iaria.

"Some people don't actually lack the skill of forming a mental map," he said, "but somewhere in the process of collecting all the pieces of the puzzle, errors accumulate, information gets lost and suddenly the map shifts."

The condition appeared to have varying degrees of severity. One of his patients' worlds would shift back and forth every minute of the day. "One moment her brain was telling her the bathroom was on her left, the next it was on her right. It literally drove her crazy," he said.

I asked Iaria what he thought of Sharon's spinning technique. He said he knew of others who seemed to be able to reset their mental map, normally by concentrating on things around them. But, to his knowledge, Sharon's technique is unique.[10]

"I have to admit I have no idea why it works," he said. "There's nothing wrong with her vestibular system—she doesn't get nauseous or have problems with her balance—but somehow shaking this system by spinning around resets her mental map."

He sighed.

"I can scan her brain, but I can't enter her mind."

Recently, Iaria has been testing his theory that developmental topographical disorientation has a genetic link.[11] Of all the people he has identified with the condition—almost two hundred of them—around 30 percent have at least one other family member affected by the problem. To confirm their suspicions, he and his colleagues have been sequencing their patients' entire genomes. They have identified a handful of potential genes that might be causing the problem. "We're really close to identifying exactly what genes are responsible," he said.

It's a huge step. This kind of research will allow doctors to sequence the genes of children whose family members suffer from the disorder and predict whether they too will develop navigation problems. While it's unlikely that they will be able to replace the broken genes anytime soon, it might be possible to intervene using brain-training exercises that help children use other parts of their brain to learn how to navigate.

"The earlier we catch it, the more likely we can do something to teach that child to learn specific navigational skills that might not develop naturally," said Iaria.

I wondered whether there was anything the rest of us can do to improve our navigational skills or if it is too late by the time we've reached adulthood. "Sure," said Iaria. "If you're in a new area you should return to one point—your home base—often as this will help you build a better mental map." He said you should also pay much more attention to your surroundings, take note of specific landmarks and think about their orientation to one another. "And don't forget to turn around or look backwards from time to time: it's a trick that animals do to make it easier to recognize their way home."

As we leave the restaurant, I ask Sharon whether her daughter, son or grandchildren have any signs of the condition.

"No, thank goodness—they are all really good at navigating," she says.

We walk a few steps in silence. Did Sharon's condition appear spontaneously, I wonder, or might it have been inherited?

"Do you reckon—" I begin.

"My mum?" Sharon guesses. "Yes, I think she must have had it. Looking back it all makes sense. She never told my dad about my condition, probably because she'd never told him about hers. She never walked us to school or picked us up from anywhere unless there were other people with us. The only time she ever left the house was when she was in the car with my father or visiting neighbors on our street. She never went anywhere by herself—never."

While it may be too late to help Sharon, just knowing there are people out there trying to understand the condition has been enough to help her turn her life around.

"I've always been silly and funny because it misdirected people away from the things I was hiding. Everyone always said, 'You're always in such a good mood.' They didn't know that I would go home at night and cry. Now I don't need to. Now all my friends know what I've got and why I have to go off and do my Wonder Woman impression when we're out."

That's not to say there aren't times when her condition is still a nuisance. Recently she got lost in a department store. She was late for a party, so in order to spin around and find her way back to her car she grabbed a pair of shorts and ran into the changing room. Only then did she realize the shorts she was carrying were for a toddler. She walked out of the changing room with her head held high.

"I just told the shop assistant, 'Sorry, they were a little small.'"

I wonder, as we drive back to Sharon's house and I recognize a few of the turnings, whether her brain is completely different from mine, or whether she is just at the far end of a navigational spectrum on which we all sit. I later asked Iaria what he thought. He said that the condition certainly exists in varying degrees of severity, but based on what we know so far, it's not possible to conclude whether Sharon is at the bottom of the spectrum or off it completely.

"Look at it this way," he said. "If you take a hundred people and relocate them in a new city, some will begin to learn their way around the area within days, others will take weeks, some will take months. After a year, all of those people will know their way around with varying degrees of confidence. But take someone with Sharon's disorder to that city and they will never be able to give you directions, not after a year, not even after a decade. They will get lost every single day. The mechanisms involved are the same, but something, somewhere, is completely different from you or me."

SHARON AND I WANDER back into her house, where she points me in the direction of the kitchen, this time to a plate of banana bread that she's made for my flight home. We stand in front of her fridge again, arguing about how many pieces of cake I can legitimately take through security. Sharon insists I should take the whole plate. I compromise and carefully wrap three slices in a piece of tinfoil. I would later receive texts and emails from her, checking that I had gotten home safely.

I tell her that I am amazed, given what she has been through, that she is so nice, so normal. I know she won't take offense at the word.

She glances back at the fridge. "The way you see me now is because of Giuseppe. I wasn't this person before I saw Giuseppe. I was still a scared little girl. I don't think I grew up and became a woman until the last ten years, really. I'm a happy person now. I realized that in order to be fulfilled I needed to learn to like myself and accept who I am."

She smiles. "Now I have Wonder Woman on my refrigerator. I'm really proud of what I am today."

Out on her doorstep I catch another glimpse of her giant lobster lawn ornament waving at me.

"I know he's awful," says Sharon, as she walks me to my car, "but I call him Louie." She looks back at the house. "If I'm lost trying to find my way around all the condos and I see Louie . . . I know I'm home."

On the plane, I look at a picture the two of us had taken together in the restaurant. Sharon's bright red hair and beaming smile shine out. From the outside you would never know there was anything strange about the way she sees the world. Yet her mountains can leap from one direction to another; the home she recognizes can change in an instant.

We're slowly creeping closer toward an idea of why this happens—just how the different cells in and around our hippocampus communicate with one another to form our internal GPS. Perhaps one day we'll know enough to fix it when it goes wrong. But in the meantime, I wonder how many others are out there like Sharon, hiding a similar secret. People who are making excuses, working out tricks, driving themselves into depression for fear of being stigmatized. All because we find it so difficult to compare objectively how we see the world.

"Beautiful, isn't it," says the man sitting next to me, pointing out of the window.

I glance down at the twinkling lights of London coming into view and smile in agreement. But I have a peculiar feeling in my stomach. A few days ago, I would have taken it for granted that we were appreciating the same thing—the dark blue curve of the River Thames, the silhouette of the Houses of Parliament. But Sharon taught me that it was entirely possible that this gentleman and I were seeing the world completely differently from each other. I look at my neighbor, thinking about whether my London resembles his.

As we approach the city, the distinctive lights of the Shard growing larger in the window, I wonder, Is there any way of finding out?

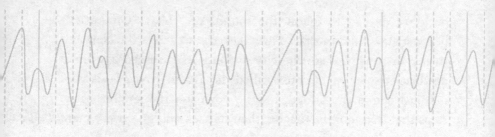

RUBÉN

Seeing Auras

I squint as the bright sunshine greets me at the end of the tunnel. My bus rumbles past the Guggenheim Museum, with its erratic curves of stone, glass and titanium gleaming brightly. Farther down the road I am joined by a gigantic dog, twenty feet tall, covered with multicolored blooms. In the distance, a needle-thin tower soars into the sky, squeezed between a Gothic church and row upon row of orange-roofed apartments.

This is Bilbao, a Spanish city that sits at the northern tip of the Iberian peninsula. It is early morning and the temperature is already climbing. I am due to meet a fellow journalist, someone who I'm hoping will help me understand how other people's worlds differ from my own. But first I have to find him.

I hop off the bus at a huge roundabout and try to figure out which of its seven exits to take. I am newly grateful for an ability to bring to mind a mental map of my surroundings, but am still finding it difficult to know which way to turn. I briefly consider attempting some Spanish to ask for directions, before deciding instead to follow the sounds of Abba's "Chiquitita" being strummed on a sitar. This takes me out across the Nervión river that runs through the city dividing its districts. My final destination—an opera house called the Teatro Arriaga—is visible just across the bridge. I settle down on the amphitheater steps that lead into the theater, then sit back and give every man who passes a lingering stare.

In the end, with his thick brown beard, black-rimmed glasses and heavy-set frame, thirty-year-old Rubén Díaz Caviedes isn't hard to miss. He turns toward me as I jump down the steps, waving awkwardly. We meet at the bottom. I hold out my hand; he ignores it.

"We do it the Spanish way," he says, giving me a kiss on either cheek.

I must look surprised. Not at the kisses but his voice.

"Ah yes, my accent," he says. "I'm told I sound like a posh Englishman."

I laugh and we chat easily as Rubén leads me toward Bilbao's old town in search of a traditional Basque breakfast: a large black coffee.

As we amble along the city's cobbled streets, Rubén explains how he traveled to Bilbao from a village just along the coast, where he works for a contemporary culture magazine. Until recently, he lived in Madrid and then Barcelona, but moved to the countryside for a better work-life balance. For the mountains and the greenery, he says. "The things you can't buy with money."

Rubén's new life is in his home town of Ruiloba, where some of his family still live. He is the oldest child of three—all boys, two and a half years apart. His childhood was a happy one, but unremarkable. Rubén was twenty-one when he first realized that he had an extraordinary brain. But to find out more, I have to ask him a question that I know he is going to detest.

"Rubén, you're going to hate me for using this word, but is it auras that you see?"

Rubén takes a deep breath.

"Yes, I suppose, if you have three hours to explain it," he says. "But if you have just a few minutes with someone and you tell them that, then people are going to think you're a magic leprechaun or—" He pauses, searching for the right word in English. "Or a twat."

In 1997, Loftur Gissurarson, an Icelandic scientist working in Reykjavik, invited ten remarkable people into his lab. All ten claimed they could see auras.

Auras are perhaps most commonly associated with religion, often seen floating around the bodies of Mary and Jesus in Christian artwork. They are alluded to in many spiritual practices as chi, prana or chakra—mystical energy centers that are said to coincide with seven major areas of the human nervous system. They have been described as a halo of color or light, or an electromagnetic field that surrounds all creatures—emanations thought to reflect health, mood and enlightenment. The majority of the scientific establishment dismisses them.

I asked Gissurarson—now a managing director at a geothermal company in Reykjavik, but once described by peers as a "jovial, convivial, pipe-smoking parapsychologist"—which camp he was in. He said that for him it was purely a matter of experimentation; he had been interested in studying the aura because at the time the phenomenon had not been put to the test, scientifically speaking.

"A number of ostensible psychics claimed to be able to see auras and I was curious to see how it would make out in a controlled situation in a lab," he said.

Paranormal activity had long captured Gissurarson's imagination. He wrote his PhD thesis on Indridi Indridason, Iceland's first and most prolific medium. In a book he later coauthored, he details the investigation of the phenomena produced by Indridason, which included making his own arm disappear, levitating and invoking the appearance of multiple voices during seances.[1] So prolific were Indridason's talents that several distinguished scientists, including Gudmundur Hannesson, a professor of medicine, twice president of the University of Iceland and a member of Parliament, studied him at close quarters. Hannesson's records were meticulous. During seances in which objects would appear to fly around the room, Hannesson would attempt to limit every conceivable means of deception. He'd place a net around the room, he would hold Indridason's feet and hands, he'd investigate the possible use of mirrors or accomplices. At the end of his study, he remarked that at almost every seance he noticed something that he considered suspicious, and at the next one would be especially vigilant on that particular point. "But," he concluded, "in spite of all [precautions], I was never able to ascertain any fraud. On the contrary, the bulk of the phenomena were, as far as I could judge, quite genuine, whatever their cause may have been."

Almost a century later, Gissurarson and his colleague Ásgeir Gunnarsson, lined up four large wooden panels in an empty room. Gunnarsson hid behind one of the panels—decided by a roll of dice—and waited for Gissurarson to bring each of their ten participants into the room, one by one. Standing at the doorway, each participant was asked to say behind which screen Gunnarsson was hidden. The researchers reasoned that the participants should be able to tell based on Gunnarsson's aura, which would be streaming out from behind the screen. Each participant repeated the experiment over and over again. The pair then invited nine people who said they had no extrasensory ability to take part in the test.

They tried hard to minimize any aspect that could potentially reveal Gunnarsson's position: the walls were covered in opaque wallpaper to prevent any reflections giving the game away; the participants were given ear defenders and listened to music between each test to stop them hearing the researcher's footsteps. Gunnarsson even took a shower just before the experiment to prevent any lingering musk from revealing his position.

The results were conclusive: neither group was able to guess which panel Gunnarsson was standing behind any better than chance—and, ironically, the control group actually did slightly better than the group that claimed they could see auras.[2]

Gissurarson wasn't alone in giving paranormal activity a chance at scientific validation. In 1964, James Randi, a renowned magician and escape artist, now best known as a tireless investigator of paranormal and pseudoscientific claims, offered $1,000 of his own money to the first person who could offer him proof of the paranormal under controlled conditions. Today, his prize lies untouched, and thanks to various donors, has grown to $1,000,000. Although hundreds of people have tried, all attempts at claiming the money have failed. Most notable was a live experiment on ABC's prime-time program *Nightline*, where a psychic, a palm reader and a tarot-card reader all had their talents put to the test. All of them failed.

"I have an open mind," said Randi, after the show had aired. "But not so open my brains fall out."

"And that," says Rubén, "is why I don't tell people I can see auras."

He and I are sitting under a large cream umbrella in a small square tucked away in Bilbao's old town. I flag down a waiter. Rubén sits forward in his chair.

"First of all," he says, quite seriously, "I don't want anybody to think that I see auras in the very conventional sense of the word, like I'm some kind of fortune-teller or that I can also read hands."

I nod.

"What in fact happens is that I perceive colors when I see people. Everyone has a distinctive color, which changes with time depending on how I know that person, or the main attributes of the person."

"Attributes?"

"Like their name, or voice, what they're wearing, and the emotion that I am feeling toward them."

"Can you physically see the color in front of you?"

"That's the most difficult thing to explain. It's not a hallucination, not something visually happening in front of you, but at the same time I'm aware that it's there. I can't avoid seeing it."

But Rubén does not have paranormal talents. He has a rare variant of synesthesia, the condition we came across in Bob's chapter that causes a blending of the senses.

FOR HUNDREDS OF YEARS, it was received wisdom that our senses travel along their own individual paths in the brain, never talking to one another directly. We see because of impulses that travel from the eye through the optic nerve to the visual cortex. We hear because air triggers electrical messages in the ear that are passed to the auditory cortex and perceived as sound. In 1812, that wisdom was challenged by Georg Tobias Ludwig Sachs, a young man born in the mountain village of St. Ruprecht in Austria, in a dissertation that he published describing his own albinism—a condition in which a lack of melanin causes a person's hair and skin to become pale white. In his essay, he also remarks upon another phenomenon in which colors appeared when he listened to music, or when he thought about numbers, days, cities or letters. He says that these concepts "introduce themselves to the mind as if a series of visible objects in dark space, formless and noticeably of different colors."[3]

It wasn't until the 1880s that Sir Francis Galton, a polymath from Birmingham in England, named Sachs's condition synesthesia, a term that comes from the Greek, meaning "joined perception." A synesthete, you remember, might experience the number five as having a pink hue, or taste strawberry at the sound of a horn. Music might be perceived as having a particular shape; months of the year might be seen as a ribbon in space. My favorite description of synesthesia comes from the Russian author Vladimir Nabokov. "The long *a* of the English alphabet . . . has for me the tint of weathered wood, but a French *a* evokes polished ebony," he says in his autobiography. "I am puzzled by my French *on* which I see as the brimming tension-surface of alcohol in a small glass . . . In the brown group, there are the rich rubbery tone of soft *g*, paler *j*, and the drab shoelace of *h*."[4]

Synesthesia is by and large a completely harmless trait, and occurs in around 4 percent of the population. Many people are synesthetes without ever realizing. Undoubtedly, these strange perceptions would once have been regarded as witchcraft. Even in the last century, synesthetes were often diagnosed with schizophrenia, or taken as drug addicts. Thankfully, the landscape has changed radically in the last few decades. Scientists no longer ask whether the condition is real, but why it occurs and whether it is beneficial.

Although the debate over the mechanisms that give rise to synesthesia is by no means settled, the increasing sophistication of imaging techniques has allowed us to compare the structure and patterns of the brain activation in synesthetes and nonsynesthetes.

At first glance, the synesthetic brain looks much like any other. It has the same tangled heap of neurons that we all possess. But upon closer inspection there may be subtle differences. As we discovered earlier, neurons in the infant brain form millions of connections that are not present in later life. As we grow and learn and experience the world, a huge amount of connections are pruned. Some small studies have suggested that people with synesthesia may have a genetic anomaly that prevents this pruning from happening in certain brain areas. As a result, synesthetes are left with pathways of communication between sensory regions of the brain that don't normally exist.

While these structural changes and coactivation of disparate regions of the brain may indeed increase a person's propensity to link different senses, it doesn't fully explain the mechanism behind synesthesia. It doesn't explain how synesthesia can be induced temporarily, after taking hallucinogenic drugs, for instance, nor does it explain the handful of cases in which people have lost synesthetic experiences while taking antidepressants.

In fact, it seems that anyone can become a synesthete. In 2014, Daniel Bor at the University of Sussex and his colleagues managed to turn thirty-three adults into temporary synesthetes in just over a month.[5] Their volunteers took part in half-hour training sessions, five days a week, in which they learned thirteen letter and color associations. By week five, many of the volunteers were reporting that they saw colored letters when they read regular black text. "When reading a sign on campus I saw all the letter E's coloured green," said one participant.

If you want to try it for yourself, you can download e-books in which certain letters always appear in specific colors. Before long, you should start seeing those letters appear in color elsewhere in the world. The effect doesn't seem to last long if you don't continue to practice. Three months after the trial finished, the volunteers' synesthesia had disappeared.

The fact that synesthesia can appear and disappear in these ways challenges the pruning theory: it's not possible for new connections to sprout suddenly and then disappear in such short time frames. The Indian-born neuroscientist Vilayanur Ramachandran poses a different theory: he and his colleagues at the University of California, San Diego, believe that synesthetes may actually have an enhancement of preexisting connections between their senses that everybody possesses.

We know that there are several areas of the brain that inhibit each other; in this way neighboring areas of the brain can be insulated from one another. There's some evidence to suggest that a chemical imbalance might reduce this inhibition, either by blocking chemicals that pass electrical messages across synapses, or by failing to produce them at all. This would not create any extra connections in the brain, but would prevent some connections from being inhibited, with the result that regions normally shut off from one another would start communicating.

If this theory proves true, you might imagine that we all have some aspects of synesthesia within us. And when we look closer, it turns out we do. Imagine you have in front of you a rounded cloud-like shape and a shape that resembles a jagged piece of shattered glass. Which would you name Bouba and which would you name Kiki? Most people would name the rounded cloud Bouba and the jagged shape Kiki. This is the most likely answer no matter whether you are an English-speaking person or not. This intriguing experiment, developed by Ramachandran, shows that while we may not see colors when listening to music, or looking at numbers, when pressed we all tend to link certain senses—like pairing high-pitched sounds with bright colors, and low tones with deeper hues. Such experiments suggest there is a non-arbitrary, inbuilt relationship between all of our senses. It suggests that synesthetes do not have a completely different brain from the rest of us; they may simply have a more extreme manifestation of what we all possess to a greater or lesser degree.[6]

IT'S NOT CLEAR how many kinds of synesthesia there are, and new types are being described all the time. In 2016, Jamie Ward at the University of Sussex discovered that some synesthetes who are fluent in sign language experience the same color they associate with written letters when the corresponding letter is signed.[7] Then there are the more unusual kinds of synesthesia: the ticker-tape synesthete, for instance, who sees words flowing out of people's mouths as they speak,[8] or the orgasm-color synesthete, who senses bright colors at the point of climax.[9]

Rubén's synesthesia is thought to be one of the rarest, in that he experiences all sorts of crossed senses. He perceives colors when he sees or hears letters, numbers, names, music, shapes or heights, when he thinks about certain ideas, and also when he feels strong emotions. This emotion-color synesthesia is what leads to his most interesting perceptions—a world of colorful auras evoked by the people around him. Sometimes the color he associates with someone is completely arbitrary, at other times certain colors are associated with specific emotions that he has toward the person.

"So does everyone have a color associated with them?" I ask Rubén, pointing at a random woman walking past our table. "Like her? What color is she?"

"No, not everyone," Rubén replies, giving her the briefest of glances. "The colors I see are primarily influenced by the sound of the person's name, what they're wearing, how I feel about them, or how attractive they are."

The colors Rubén sees often are blue, gray, red, yellow and orange.

"For instance, if I like someone sexually they would be red," he says. "It won't matter about the voice, just the look, because that's the first thing you think about a person. It's not just people but also music, paintings and buildings. Things I like always tend to make me perceive some kind of red."

People who look dirty or sick, on the other hand, will normally be perceived by Rubén as having a green aura, whereas those who are optimistic and happy are purple.

"If I don't like someone, their color will probably be yellow. Yellow is the color I associate with acidic flavors and it's also the color of people with bad manners, who are rude or have a kind of attitude. So if a person acts like that, he or she becomes yellow."

Rubén doesn't always have an explanation for why certain colors are associated with certain people. One of his brothers is pale orange, the other is gray, and his mother is gray-blue. He has no idea why. Likewise, his father is brown. Brown is the color Rubén normally associates with the elderly and people who are uninteresting to him, yet his father is neither of those things.

"In their case, it doesn't have anything to do with emotions. It's more to do with their identity and how their voice sounds."

Sometimes people's color changes, he says, sipping his coffee. "I had a boyfriend up until a few years ago and the first time we met, I remember thinking he was this bright red. But he had this amazing voice and these blue, almost green, eyes—and those two things, the color of his voice and the color of his eyes, were so distinctive that they mixed and that became his color. It was this pale gray. No one else had that color."

THE RELATIONSHIP BETWEEN COLOR and emotion is well established in the animal kingdom. Female animals often use the color red to signal hormonal changes in their body associated with fertility, for example. Certain male primates show red following a surge of testosterone in their bloodstream, due to aggression or as a show of dominance. Testosterone suppresses the immune system, so the flush of red tells any females that the male must be in good health to cope with such deficits.

There's plenty of research suggesting that color affects us, too. Consider this simple yet remarkable social experiment, conducted in 2010 by Daniela Kayser, a psychologist at the University of Rochester, New York. Kayser wondered whether a lady in red really was more alluring, so she and her colleagues asked several men to have a conversation with a woman who was wearing either a red or green shirt. Men who spoke to the woman while she was wearing a red shirt asked her more intimate questions than those who spoke to her while she was wearing green. In another experiment, men sat closer to a woman and classed her as more attractive when she was wearing a red shirt than when she was wearing an identical shirt in other colors.[10]

The results certainly fit with our mainstream ideals of red being linked to a woman's allure, passion and fertility. But men, take note. Over a series of seven experiments, Kayser's colleague Andrew Elliot demonstrated that women also perceive men to be more attractive, more desirable and considerably more likable when they are wearing red clothing.

Colors also influence other aspects of our behavior. In humans, aggression and dominance are associated with reddening of the face due to increased blood flow—perhaps that is why we refer to "seeing red" when angered. Evolutionary anthropologists at Durham University and the University of Plymouth wondered whether wearing a red shirt might exploit our innate response to the color red and so influence the outcomes of sporting contests. They studied fifty-five years' worth of English football league results, and found that teams whose home colors were red won 2 percent more often than teams who wore blue or white, and 3 percent more often than those who wore yellow or orange.[11]

In fact, across a range of sports, wearing red is consistently associated with a higher probability of winning. Robert Barton, who worked on the football study, also analyzed the results from four combat sports in the 2004 Olympic Games. Despite the athletes having been randomly assigned red or blue outfits in which to compete, those who wore red won 55 percent of fights.[12]

Barton says it's not completely clear why this happens— whether the color red affects the wearer, the perceiver or the referee. "There is some evidence that wearing red increases feelings of confidence and hormone levels," he says. There's also evidence that the color red can affect the judgments of referees, and that people associate the color with dominance, aggression and anger, which might have subtle effects on an opponent's performance.

"It's an interesting question as to why in so many cultures red becomes associated with the same kind of things," says Barton. "It suggests there is some universality about it—whether that is a direct reflection of an evolutionary heritage or something else that makes red so salient."

Despite this uncertainty, it seems that colors do unwittingly affect all of us day by day. If Ramachandran's theory that we have a nonarbitrary, inbuilt relationship between the senses is correct, we might all have the anatomical connections in place to link emotions and colors; it's just that for most of the time we inhibit these pathways to varying degrees. Perhaps this is why the color red influences our behavior in subtle but provocative ways. At the very least, it might give you a few ideas for what to wear on a first date.[13]

An accordion player edges nearer to our table, so we decide to make a move. I pay for our coffees as Rubén resumes his tale, recollecting some of the things that happened to him as a child, that in retrospect seem relevant to his synesthesia.

"I've always hated my hands," he says, holding them up to my face. "They're like giant baby hands."

I suppress a smile. They are very much like giant baby hands—thick stubby fingers on squishy circular palms.

"The weird thing is that I used my right hand to draw and I was quite good at drawing so I started to like my right hand, but I still hated my left. Whenever I would imagine my hands, I would see the right one as this beefy Conan character and the other one as a little evil character. I'm pretty sure that had something to do with my brain making strong visual representations based on my emotions."

Other strange things happened as Rubén was growing up. There was a period when he would perceive a woman dancing when he looked at certain things—his teachers, his friends, even his dog. He couldn't avoid seeing it.

These strange perceptions, which had started off as flickers of dancing women and pantomime hands, had, by the time he was a teenager, solidified themselves as auras.

"There was clearly always something weird happening in my brain," he says.

As we wander away from the hustle of the old town, through a maze of small side streets in search of something to eat, I wonder whether the colors that Rubén perceives give him any kind of special insight into his emotions.

"Do you ever see someone and notice that you're perceiving a red aura, say, and think, 'Oh, I must fancy him'?" I ask.

Rubén laughs.

"No, it doesn't work like that. When it's happening, the color is the effect of the emotion. The order of events are person, emotion, then color. So I already know what emotion I'm feeling."

He pauses.

"Except, actually, sometimes it's the color, then the emotion and the person."

He searches the crowd for a moment, and points at a passing tourist.

"When your emotions are linked with colors, it tends to go both ways. So I might see someone in bright-red trousers and because I associate red with love or attractiveness, I could be aroused by that person or think better of them. You know it's something stupid and irrational, but it tends to get inside your brain because you can't ignore it. You have to tell yourself, 'This person isn't nice just because they're wearing red.'"

"So might you think someone was nasty because they were wearing a color that you associate with rudeness?" I say, glancing down at my blue dress, wracking my brain to remember what emotion Rubén associates with that color.

"Exactly," he says. "If they were wearing something very yellow or I see someone with a green aura because that's the color I associate with their voice, I might feel tempted to think that person is less nice because their greenness makes me feel a certain way."

"Isn't that kind of annoying?"

"It could be, but the important thing is that I'm perfectly aware that it is irrational. I know these feelings are stupid, I just have to fight them. None of it is real."

"Do you think you've been this way ever since you were born?"

Rubén stops and thinks for a moment. "I have the sense that I could always see colors associated with people, but when you don't experience anything different, you don't realize it's unusual."

In fact, it wasn't until 2005 that Rubén became aware of his synesthesia at all. He was hanging out with a friend who was studying psychology at the University of Granada. She told him that she was involved in an investigation into synesthesia. He had never heard of the word, so she explained what it was.

Like many others in the past, Rubén didn't understand why it was worth investigating.

"I was like yeah, yeah, what of it?" said Rubén. "That's completely normal!"

His friend was surprised and told him that she thought he might be a synesthete.

"Then suddenly she went completely white," says Rubén. "She remembered that I was color blind."

In order to see the kaleidoscope of color in our world, we use specialized cells in the retina called photoreceptors. These absorb light and convert it into electrical signals. There are two kinds of photoreceptors, rods and cones. Rods help us to see in dim light but are not sensitive to color. Cones, on the other hand, respond strongly to red, green or blue. When wavelengths of light hit our cones they respond optimally to their favored color or to a lesser extent to wavelengths of light that are close to their favored color. For example, cones that favor red light also respond to orange and slightly to yellow, but don't respond at all to green and blue. The combination of activity from all three types of photoreceptor is sent to an area of the visual cortex called V4, where it is interpreted as the many shades of color that make up our Technicolor world.

However, for people who are color blind, like Rubén, some of these photoreceptors are deficient, resulting in the loss of a whole spectrum of color. Rubén has a common form of color blindness that makes it difficult to tell the difference between colors that have some aspect of red or green in them.

"I can tell the difference between a lettuce green and a lipstick red, but colors in between like purple and some blues and oranges get mixed up," he says.

Being color blind has given Rubén a bit of a complex about colors and was the reason, he reckons, that he never really let himself think much about the colors he perceived around people, letters and buildings.

"What bothered you so much?" I ask.

"You know when you're in kindergarten and coloring a picture and you need a crayon?"

I nod.

"Well, I would be coloring a picture of a person and I'd ask for the pink crayon. The other kids would give me another color and wait for me to color the face blue. They'd do it as a joke, but I didn't like it. You're only three, your only job is to learn the colors and you're not able to do it. It's not nice, you know?"

Rubén remembers one time when he painted a horse. It was a decent horse, he says, but when the teacher came over to take a look, she was really, *really* impressed. Then she asked him why it was green.

"I was so embarrassed about it being green," says Rubén. "I just said, 'Because it's nicer.'"

That particular teacher, not knowing Rubén was color blind, was reminded of a printmaker called Franz Marc who painted a famous picture of blue horses against red hills. Marc used colors to express strong emotional meaning or purpose. Rubén's teacher wondered whether she was seeing the start of something quite profound in this young boy. She was so impressed with his artwork that she invited his parents to the school to discuss his future.

"She told them I had been doing these wonderful colorful paintings. She thought I was a genius," says Rubén. "My mum was like, 'Er, no, he's really not!'"

But Rubén's teacher was right—there was something special about him.

AFTER RUBÉN'S FRIEND RECOVERED from her shock, she took him to the University of Granada to meet her supervisor, Emilio Gómez, a cognitive psychologist.

"He was quite emotional when we first met," says Rubén. "I don't think anyone thought a color-blind synesthete could exist."

The reason that Gómez was so excited to meet Rubén was that he believed he might be able to offer a novel insight into the question I'd first considered on the plane back from Sharon's: Does my world look like yours?

Scientists call this concept qualia. To understand its meaning, imagine I'm an alien visiting Earth from another planet. I ask you, What do you see when you look at that red apple over there? You could tell me all of the physiological mechanisms that occur when you look at the apple. You could explain how wavelengths of light hit your eyeballs and send signals toward areas of the brain that process color. You could tell me about all the other things that look red, or how it makes you feel. But your description leaves out something completely ineffable: your actual perception of what red is. We are fundamentally unable to transfer the experience of our world to other people.

What we are starting to realize, though, is that we don't always see things in the same way. Never was this more obvious than in February 2015, when the world stumbled upon a certain blue and black dress. Or maybe, like me, you thought it was white and gold. In case you missed the year's biggest debate, it surrounded a simple photo of a perfectly nice, blue-and-black-striped bodycon dress. If you haven't seen it, I urge you to google it immediately. The photo was uploaded by twenty-one-year-old Caitlin McNeill, an aspiring singer from Scotland, after some of her friends swore the photo showed a white and gold dress. Social media went into overdrive, people in the blue-and-black camp could not understand why so many of their friends were seeing white and gold. The television host Ellen DeGeneres tweeted, "From this day on, the world will be divided into two people. Blue & black or white & gold."

Scientists rushed to cobble together an explanation.[14] When light hits an object, they said, some of it is absorbed and some of it is reflected. The wavelengths of the reflected light determine the color we see. The lightwaves hit the retina at the back of the eye and activate our cones. A combination of activity from the cones is sent to the visual cortex in the brain, which processes all sorts of visual aspects such as movement and object recognition, before eventually producing the perception of color. So far, so good. However, those wavelengths of light are actually a product of whatever colors of light are currently around you, reflecting off the object you're viewing. The light that illuminates our world changes throughout the day, from the pinkish light of dawn to the bright white neon lights of your office and everything in between. Without you being aware of it, your brain considers what color light is bouncing off the object in your vision and makes certain adjustments to compensate. This mechanism allows you to walk through shadows, or in and out of a brightly lit room, while keeping the colors of your world stationary.

Scientists decided that The Dress must lie on some kind of perceptual boundary. In other words, it wasn't clear in what light it was taken. Which meant that some people's brains adjusted for a blueish light and ended up seeing the dress as white and gold, while others—correctly, as it happens—discounted the gold end of the spectrum and ended up seeing it as blue and black.

I find it difficult to look at The Dress and not be just a little alarmed, because it reveals something of qualia that we so often take for granted: The colors I see are not always the colors that you see.

And for Gómez, Rubén's color blindness together with his synesthesia was the perfect way of gaining a unique insight into this unexplainable matter.

But first he had to prove that Rubén was telling the truth.

Rubén stared at the hundredth image that day and pointed to a color chart that correlated with the aura the image produced. It was 2010, and Gómez had asked him to complete this task so that he had a record of what color auras Rubén associated with certain images of faces, animals, letters and numbers. There were so many images it would have been impossible for Rubén to memorize each one.

A month later Gómez surprised Rubén by asking him to repeat the task. His answers matched almost 100 percent of the time.

Satisfied that that test had been passed, Gómez's team devised a personalized Stroop test for Rubén. The original version of this test asks participants to name the color of a word, not what that word says. For example, if the word "red" was written in blue ink, you would say "blue." People find this task much easier when the word and color of the ink match. We can read words more quickly than we can process colors, so when the two are incongruent, it makes the brain stumble, meaning it takes longer to reach the correct answer.

Gómez's team tweaked this test in a number of different ways to evaluate Rubén's claims. In one, they asked him to state whether a number was odd or even, but presented these numbers in colored inks. The inks either matched or contrasted with the color of the aura that Rubén perceived when he read the number.

Rubén's reaction times were quicker when the color of the number matched the color of the aura it produced. We're not talking seconds here, but mere fractions of a second quicker every time—something that would be impossible to fake consistently. Since the colors of the numbers are arbitrary for people who don't see auras, their reaction times were similar across the board.

Once he was convinced that Rubén really was telling the truth, Gómez began to devise a way of testing whether Rubén's auras affected his behavior. To do this objectively, he needed to test a behavior over which Rubén had no conscious control: his heart rate.

Gómez showed that Rubén's heart rate rose ever so slightly when looking at pictures where there was an incongruence between the aura they produced and the content of the picture—for instance when an attractive man was wearing green. The feelings of attraction were at odds with the emotions that the green clothes produced. It was an image that was described by Rubén as being "emotionally inconsistent."

In contrast, the heart rates of people who don't have emotion-color synesthesia and who took the same test didn't fluctuate at all.[15]

"It seemed reasonable to conclude," said Gómez, "that Rubén's bodily reactions were a unique result of his qualia, or experience of color."

Although it doesn't tell us exactly what Rubén sees, it does answer my question of whether our worlds look alike. The answer is no.

RUBÉN AND I ARE DISCUSSING this complex concept when he says something that makes me stop quite literally in the middle of the street. Although he cannot tell the difference between shades of green in real life, he says, he sees a number of shades of green aura. "I only have one type of red in my mind, the red I see in real life, but I have a number of greens—not just one."

I am startled by his admission. It suggests that Rubén can see colors in his mind that don't exist for him in real life. He likens it to seeing someone in a dream: "You can't see their face but you know who it is, despite how they look."

There are other qualities of his auras that don't exist in real life. The colors are textured, translucent, he says. "Some have a sparkly, glitteriness about them."

It turns out that there is only one other person known to have this extraordinary, and rare, combination of synesthesia and color blindness. He is Spike Jahan and he is a student of Ramachandran. Jahan approached Ramachandran shortly after he had attended a lecture on synesthesia. He told Ramachandran that he was color blind and had trouble distinguishing reds, greens, browns and oranges. He also had number-color synesthesia. However, the colors Jahan saw in his mind were tinged with colors that he had never seen in the real world. He called them "Martian colors."

I asked Ramachandran to explain this mysterious phenomenon to me. He said that Jahan has deficiencies in his cones, meaning he is unable to see certain colors in real life. Yet those deficiencies are in his eyes and not in his brain. The part of the brain that processes color is perfectly normal. Somehow, when Jahan looks at a number, the shape of the number is processed normally, but then crossed wires activate the color area in his visual cortex, which triggers the sensation of colors that he is unable to see in the real world.[16]

Although Ramachandran has not studied Rubén, he said he would guess that a similar thing is happening in his brain. Perhaps parts of his brain that deal with emotions are able to stimulate areas of the visual cortex, which enables him to perceive shades of green that he is unable to perceive in reality.

Despite these being single-case studies, they hint at yet another mysterious aspect of qualia. Jahan and Rubén's Martian colors suggest that what you call red is not determined purely by light-waves or the photoreceptors in your eyes, but is an innate concept produced by the activation of certain color regions in your brain. It suggests that color doesn't have to be activated via visual stimulation, but is an experience that can be a property of shapes, sounds or emotions. Perhaps in the future, says Ramachandran, we'll be able to stimulate these color regions alone to discover what strange experience they might provoke—a feeling of redness, a sound or taste of redness, a weird mass of redness unattached to a particular object? Maybe then, he says, we'll be able to find out exactly what "red" is.

Distracted by these thoughts, Rubén and I somehow end up sitting outside a tourist trap selling bad paella. As we push food around our plates, I ask Rubén how his auras affect him day by day.

He says he is intrigued about what is happening in his brain and more than happy to take part in experiments, but generally he tries to ignore them.

"I don't really think about them much on a daily basis." Rubén frowns, taking a drag on an e-cigarette. "I think it's because if you stop and think about it too much, you feel very stupid."

If it were me, I say, I might think about using my auras to make myself feel better about myself. "Maybe wear something red because it makes you feel attractive?"

He shakes his head. "You might feel tempted to wear a certain kind of clothing because of the emotion associated with it. You could do that, but that would be stupid because it's a language nobody else speaks."

I tell him about Daniela Kayser's experiments with men and women wearing red shirts, and that in fact we might all speak his language to some degree. "That's interesting," he says. "It's quite comforting to know I'm not completely weird."

He glances down at his black top.

"I don't actually own any red T-shirts. I mainly wear black and white. I've never thought of it before, but perhaps I wear those colors more because I don't get much emotion with black or white." He smiles and looks up. "Or maybe because they're just more flattering for a big boy like me."

As I signal for the bill, Rubén asks me a question. "Do you want to know what color I see myself?"

"Yes!" I hadn't considered that his auras would extend to his own reflection.

He looks slightly embarrassed. "Red," he says. "I know that sounds like I love myself or something, all very Freudian. But I think it's because I like myself, I am happy with myself."

Rubén kindly offers to drive me to the airport. As we walk to the car I find myself contemplating the scenery around us, the deep blue of the Nervión river and the dark green mountains in the background. If it is true that colors are innate, that they can be triggered by any kind of sense, that we are all synesthetes to some extent, then surely we don't have to have sensory anomalies as extreme as Rubén's to experience the world ever so slightly differently from one another. Perhaps the only aspect of qualia that we can ever be sure of is that your red is never going to be exactly the same as mine. It gives me an excited jolt deep in my stomach. It was fun to think that my world might look completely unique. That there was something about the world that was mine and mine only.

As Rubén and I cross a bridge and amble along a tow path that runs alongside the river, my thoughts turn to a question I've been wanting to ask all day.

"Rubén?"

"Yes?"

"Do I have an aura?"

It feels strange asking. I know his colors don't always represent a specific emotion, but I am still hoping that I won't be green.

He stops walking and stares at me, tipping his head to the side. "Yes, yours is a kind of orangey color."

"Ah, phew."

"I think I see this color on you because it is the color of the particular sound of your voice. And also, if I think of you, the beginning of you would be translucent—as that's the color of the beginning of your name, and then it turns into this orangey color. So you're like a light orange with a hint of translu—"

He is interrupted by a topless jogger running past, wearing tiny blue shorts.

Rubén gazes after the lanky figure, sweat dripping off the hairs on his back. He glances at me out of the corner of his eye, shakes his head and grins.

"Definitely not red."

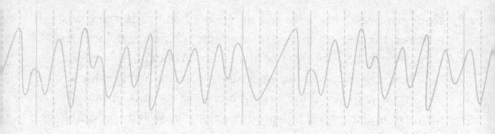

TOMMY

Switching Personalities

In 2000, schoolteacher Luke found himself in a terrible situation. He had developed an increasing interest in child pornography. He began acquiring a collection of pornographic magazines and photographs off the Internet that focused on children and adolescents. He started soliciting prostitutes at massage parlors. He went to great lengths to conceal his activities—he knew they were completely unacceptable, but later said that the "pleasure principle" overrode his attempts at restraint. It was only after he started making subtle sexual advances toward his stepdaughter, who subsequently informed his wife, that Luke's pedophilia was discovered and he was arrested for child molestation.

The judge told Luke that he had to attend a twelve-step program for sex addiction or face jail. Luke chose the program but was expelled after repeatedly asking the nursing staff for sexual favors. The evening before his sentencing, Luke took himself to the University of Virginia Hospital. He said he had a headache and was scared he was going to rape his landlady.

Doctors scanned his brain and dropped a bombshell: there was a tumor the size of an egg in his right orbitofrontal lobe, an area toward the very front of the brain. Although this region can show considerable variation among people, growing evidence suggests it is involved in working out what rewards or punishment we might receive for specific actions, as well as providing us with drive, motivation and judgment.

Surgeons removed the tumor and, just like that, Luke's pedophilia disappeared. After seven months, he was deemed no risk to the public and returned home to live with his family. A few years later, Luke's pedophilic urges recurred—this time he went straight to the hospital. Scans showed his tumor had grown back in the same place. With its removal, his personality once again returned to normal.[1]

Few demonstrations of the fragile nature of our personality are as remarkable as Luke's. But changes in personality are not rare. More than five million Americans are living with Alzheimer's, a disease that can severely affect a person's personality; every three and a half minutes in the UK someone has a stroke, which can induce temporary or permanent changes in mood, values and impulsiveness. We tend to think of our personality as something that is steadfast and strong, but in truth it can rapidly desert us.

A few years before I started this book, I struck up an online friendship with a person who had experienced not one but two completely different personalities. A man called Tommy McHugh, whose behavior, thoughts and motivations had changed dramatically after a burst blood vessel damaged his brain. But I'd only known one side of him—the person he was after his stroke. So I decided to visit his daughter to find out more about where our personality comes from, and what it is like to experience two in one lifetime.

Tommy's story starts with a potato. First it was just a few plants that showed signs of gray-green spots. Then the spots became bigger, turned brown and rough and firm. Soon, the fungus had spread to nearby crops, finally ravaging whole fields. The great potato famine, as it became known, sparked a period of mass starvation and disease in Ireland that killed more than a million people.

More than a million others emigrated. Between 1845 and 1852 several thousand families settled just across the Irish Sea in Liverpool. But they were far from welcome. Contempt for the Irish was vocalized by Benjamin Disraeli, who served as prime minister several years after the famine, when he called them a "wild, reckless, indolent, uncertain and superstitious race" that has "no sympathy with the English character." Their ideal of human felicity, he said, is an "alternation of clannish broils and coarse idolatry." As a result of this prejudice, many Irish immigrants faced daily persecution, discrimination and physical attacks.

Although Tommy McHugh was born one hundred years after the famine, discrimination was still rife in Liverpool. Having a strong Liverpudlian accent did little to hide the fact that Tommy was from a poor Irish family. He quickly learned how to defend himself against the mental and physical abuse thrown at him at school. As did his brothers and sisters—all twelve of them.

"We never let any taunting go unpunished," said Tommy, when we first spoke over the phone. "I learned to fight with my fists from a very early age."

He also learned to hide his emotions—a lesson that came from his dad, whom Tommy described as a hard worker but a drinker, "never coming home with as much money as he should."

As a result, Tommy struggled to stay on the right tracks.

"It was a tough life. I was a naughty boy. In and out of school all the time. Drugs, stealing, fighting. I did it all."

"DAD WOULD TELL US how he had to steal people's shoes because he didn't have any of his own," says Tommy's daughter, Shillo.

I am at her house in Buckinghamshire, just outside London. It is lunchtime, and black clouds are wandering all over the county, darkening the skies. We sit at the kitchen table, facing the living room, where Shillo's young son Issac is constructing a large wooden rail track. Rapid bursts of color flash from the cartoons playing on the TV—a treat for Issac, part of a deal made earlier in return for his patience while I talk with his mum.

I ask Shillo about her father. I want to know what he was like as a dad, what she remembers about his past, what kind of person he used to be.

"When he was young, it was a case of survival to a huge degree," she says. "Dad and the others stole for what they needed. Very few of his brothers weren't in prison at some point or another. He was never emotional. Never."

Tommy became a building contractor, married his childhood sweetheart and along came Shillo and her brother, Scott.

Despite his lack of formal education, Tommy loved to read. When Shillo was little, he read her *The Lord of the Rings*. As a teenager, Shillo re-read all three volumes. She remembers feeling disappointed when she found that much of the story she had loved was missing.

"I realized that Dad must have made up a load of the chapters. I was like 'What about when Bilbo did this, or when he met that person?'"

When times were good, they were really, really good, Shillo says. "He'd be fun and entertaining and the dad that all your friends said they wished they had."

But then there would be what she refers to as "incredibly dark times." Tommy struggled with anger and aggression, and often took hard drugs, including heroin.

"You never knew which dad you were going to get. He could be violent if he was drunk, there would be occasions when we'd be packed up by my mum and taken away and he'd threaten her and say, 'If you leave me, I'll find you and I'll burn the house down.'"

Shillo's voice softens.

"But then he was always good at getting everything back on track, being good and wonderful and talking with you and having lots of fun. It would be like that for a while, everything would be great. And then the darkness would return."

Differences between personalities are clear to see in real life, but incredibly difficult to study objectively. Many scientists attempt the task by defining our personality in terms of traits, or patterns of behavior, thoughts and emotions that are relatively stable over time. The extraordinary variety of personality traits is commonly broken down into the so-called Big Five, namely: openness, conscientiousness, extroversion, agreeableness and neuroticism.

Openness refers to having a general curiosity and willingness to take on new experiences, information and ideas. Conscientiousness is the ability to regulate your impulses, plan your life and display self-discipline. Extroverts tend to partake in a breadth of activities, are talkative, assertive and happy to be the center of attention. If you have a high level of agreeableness, you value getting along with others, so you may be more willing to compromise; you are kind, generous and considerate. Finally, neuroticism is a measure of how anxious you are, and your general tendency to experience negative emotions. The degree to which each of these traits exist in an individual is thought to predict their personality.

But what causes us to express these traits? Is our personality a product of our genes or our environment? To find out, we need to travel to Ohio, once the home of two very unusual brothers.

Jim Lewis and Jim Springer were identical twins, separated just weeks after they were born, renamed by their adoptive parents and raised apart. When they were reunited thirty-nine years later, they discovered that their name was not all they had in common. They both suffered from tension headaches, bit their nails, worked in law enforcement, enjoyed woodworking, smoked Salem cigarettes, and drove the same model car. They holidayed at the same beach in Florida, and both married women named Linda, only to divorce and remarry women named Betty. Both men have sons, called James Alan Lewis and James Allan Springer. They both even called their family dog Toy.

Was it just a coincidence? The behavioral geneticist and evolutionary psychologist Nancy Segal at California State University, Fullerton, says there is more to it than that. The Jim twins story was the catalyst for a groundbreaking experiment called the Minnesota Study of Twins Reared Apart, initiated in 1979. Over the course of twenty years, researchers at the University of Minnesota followed the lives of twins who were separated at birth. They studied 137 pairs of twins in all—81 pairs of identical twins who had come from one egg that had split in two, and 56 pairs of fraternal twins who arose from two different eggs.

Several researchers, including Segal, analyzed data from the study, along with data from a separate registry of twins who were raised together. They came to a remarkable conclusion: identical twins raised apart were as identical in personality as those raised together. Some traits, including leadership, obedience to authority, resilience to stress and fearfulness were more than 50 percent influenced by their genes.[2]

The results imply that a child who is genetically predisposed to being shy may become more or less shy through their upbringing but is unlikely to become an incredibly extroverted adult.

"It was really surprising," said Segal, when I asked her whether she had predicted such a dramatic result. "We expected to see more differences between the identical twins who were raised apart, but we just couldn't find them."

These studies have had their fair share of critics—one long-held complaint is that twins may have similar personalities merely because they look so much alike, which would tend to elicit the same sorts of behavior from other people.

In 2013, Segal found a way to test this theory. If true that physical appearance triggers a specific treatment by others, then the personality similarities of doppelgängers—people who look alike but have different genes—should be as similar as identical twins.

To find out, Segal recruited twenty-three pairs of doppelgängers from a project by the French Canadian photographer François Brunelle, who had been creating black-and-white portraits of doppelgängers for years. Each participant was given a questionnaire that assessed their personality using the Big Five traits, as well as other characteristics, such as degree of self-esteem. The result? Doppelgängers did not share significant personality traits, and were significantly less alike in personality than both identical and nonidentical twins raised together or apart.[3]

So is it their shared genetic history that explains the number of similarities between the Jim twins? "It's not that there's a specific gene that will make us want to holiday at the same beach," says Segal, "but why do you choose a beach holiday? It might be because you react badly to the cold, or that you are very sociable, and so like crowded places. These things are partly a function of your genetic tendencies. Taken collectively, they might explain why you're more likely to choose one holiday destination over all the others."

In the nature-nurture debate, though, nurture still has a fundamental role to play. One of the most impressive demonstrations of the environment's influence on personality came from a series of studies in the 1990s by Robert Plomin at King's College London and his colleagues, who showed that unique life experiences had the greatest influence on well-being and depression in identical and nonidentical twins.[4]

None of these studies are perfect, but the results suggest that we don't inherit a blueprint on which our personalities are permanently stamped. Our genes may predispose us to certain paths, but our personalities can be shaped by our environment over a lifetime.

And sometimes—they can change overnight.

Tommy had been suffering from a headache. It just wouldn't go away. But that wasn't unusual—he was often found with a belt tied around his head, he told me, trying to alleviate a migraine that would keep him locked up for weeks.

He was on the toilet, reading the paper, when it happened.

"I suddenly felt an explosion in the left side of my head and ended up on the floor. I think the only thing that kept me conscious was not wanting to be found with my pants down. I stood and pulled my trousers up, then the other side of my head went bang."

Tommy had experienced a subarachnoid hemorrhage due to a ruptured aneurysm. Burst blood vessels had sent blood squirting in and around his brain. He was discovered by Jan, his second wife, and rushed to the hospital, where surgeons operated on him for eleven hours. Doctors warned Shillo and her family that it might be a long time before he woke up.

"Once," says Shillo, "Dad had to go work in Saudi Arabia for a while. I must have been about three or four. He'd write to me all the time, every two or three days. When I was thirteen I was looking at the envelopes from those letters and I noticed that all the stamps were from Liverpool. I asked Mum why, and she said it was because Dad would give them to people to take back to England and then they'd post them from there."

Although doctors had managed to stem the bleed in Tommy's brain, there was no doubt that some damage had been done. His doctors were pleased to see him sitting up in bed just days after the operation. Unfortunately, there was an unexpected complication.

"As soon as I woke up, I knew immediately that something was different," said Tommy. "My mind had changed totally and dramatically."

"I was sixteen when I found out about the prison sentence," says Shillo. "I'd just started college and this girl was saying they lived next door to a McHugh. Everyone knew a McHugh because there were so many of them. Anyway, she said that all of them had been in prison, one had even been in for murder. I went home and asked what was going on and found out Dad hadn't been in Saudi Arabia. He'd been in prison."

A single fingerprint had been discovered on a fraudulent check—it was Tommy's. He claimed this wasn't possible because the finger it came from had been severed in a fight that he'd had when he was sixteen. Ever since, that particular finger had stuck up at a weird angle, unable to bend.

"He always denied the charge," says Shillo. She hesitates, and I can't work out whether she believes it or not.

"He said that finger could never have touched a check. But at the same time he had done so much bad stuff that he said they were always going to catch him on something, so it might as well be this."

I asked Tommy to describe how he felt when he woke up from his surgery.

"I was totally emotional for a start," he said. "And I just couldn't imagine hurting a fly."

Tommy looked around his hospital ward and through the window to the grounds outside. "I could see the beauty in everything. I had all these thoughts in my head that I'd never had before. I suddenly had these emotions and cares and worries. I could taste the femininity inside of me."

"It was like two different people," says Shillo. "He was incredibly emotional, he'd cry at the drop of a hat, he'd be sad, he'd be happy. The person he was before seemed to have just disappeared."

Tommy's sudden appreciation of the beauty of the world and

his new emotional compass weren't the only changes he was experiencing. Looking out of his hospital window, he saw a tree sprouting numbers.

I thought I must have misheard. "You saw numbers out there on the tree?" I asked.

"No, they were numbers in my mind," he said. "The numbers three, six and nine—then I couldn't stop talking in rhyme."

"Rhyme?"

"Yeah, I had this urge to rhyme all the time." He laughed. "There I go again. I was spouting poetry left, right and center. New stuff, old stuff, I wander lonely as a cloud. I can recite poetry upside down, sideways, on an angle—you name it, I'll say it."

A month passed and Tommy was well enough to go home. It wasn't wholly obvious to his doctors what was wrong with him. Although they knew that his hemorrhage had likely damaged parts of his brain, during his emergency surgery they'd had to insert a metal clip to stop the bleeding, which meant he wasn't able to have any more scans to pinpoint the exact regions of damage.

Tommy described his brain as having gone into overdrive. "If I go for a walk inside my brain, I see all this information," he said. "Angles, languages, structures, mathematics, wild colorful pictures. Everything I look at sparks six memories or emotions or smells, they're each spinning in my mind—just for a moment—and then it's like one of those thoughts crashes against another and that sparks six different thoughts, and then the corners of those thoughts touch and create six more. I'm constantly bombarded with patterns and details and information and faces. It's like walking inside a corridor of endless, endless information.

"My brain is like bees in a hive," he said, barely taking a breath. "In the middle, all you see are honeycomb cells covered in plastic wrap. When you stroke those little honeycomb cells, lots of other cells break out from it, like a lightning flash touching a brain cell. And from that cell comes a volcano, emitting Fairy Liquid bubbles with billions and billions of images. They're pouring out like Mount Etna, they never stop. Each of these bubbles contains another million images. That's a split second in my mind. I feel like I've been shown just how endless the brain is. It's inconceivable, we use such a tiny percentage of it."

I tried to interrupt, but he went on.

"My brain is filled with endless details but I'm too uneducated to understand all the information that's popping up inside there. It's telling me there are all these different languages, all this knowledge, pinpricks of it, microscopic hints of it all, so that if I wanted to use it, it would be there for me to use. I feel like I could talk Italian if the right thing triggered it: it's all within me. I feel like we've all got sweeping talents in our brain but we don't know they are there because we've never been forced to use them. That's my vision of what I see in my brain."

More explanations followed. I was having trouble getting a word in edgewise. As had become clear within five minutes of speaking with him, this constant bombardment of thoughts and associations was reflected in Tommy's speech. His mind traveled rapidly from one concept to the next, his thoughts turned on a dime.

Tommy would often send me pages of emails with things that he had forgotten to tell me in our telephone conversations. Some of them were written normally, some were written as verse.

The descriptions that he would weave into his speech were sometimes fanciful, sometimes insightful. He regularly came across as wise, yet when I listened back to recordings of our conversations, his metaphors were often ambiguous and disjointed.

"I feel like I've been unplugged from the Matrix," he said, one day. "Suddenly I've been disconnected from the old life that was showing me things just as someone in charge wanted me to see them.

"I'm glad I'm a bit stupid, Helen," he said. "Otherwise I would see reality far too much."

Understandably, Tommy's family was finding his endless prose, philosophical musings and gentler manner a little difficult to get used to.

"He was incredibly different," says Shillo. "His whole world was tipped upside down."

Everyone assumed that, given enough time to recover, he would change back to the person he had once been, that they'd see a glimpse of the darkness again, but it never happened.

Not everyone welcomed the new Tommy. Some people wanted him to be the person he was before, some embraced the person he had become but drifted away, finding they had little in common, others feared it was just an act.

"A lot of his brothers wanted him to turn back to who he was," Shillo says. "One in particular always tried to pull him back into trouble."

Tommy's first wife, Shillo's mum, also found it hard to accept the new Tommy. "Even ten years after the strokes, my mum still didn't believe he could really have changed," says Shillo. "She still believed that bad person was in there somewhere."

How is it then that our personality can change so dramatically? To understand that, first we have to drop a notion consistently repeated in popular culture, that we are either left- or right-brained. This theory was born in the winter of 1962, when William Jenkins, a military veteran, was being prepped for surgery at White Memorial Medical Center in Los Angeles.

The renowned neuroscientist Roger Sperry was preparing to split Jenkins's brain in half. Jenkins had been hurt in an explosion during the Second World War, and ever since had experienced up to ten seizures a day. Sperry believed that by cutting through the corpus callosum—the structure that connects the two sides of the brain—he could relieve Jenkins of his seizures. Experiments in animals suggested that this would not damage his cognition, that each side of the brain could work independently from the other.

The surgery was a success and Jenkins's cognition was, on the surface, unchanged. But further experiments on him and other split-brain patients were revealing. They proved, for instance, that the left-hand side of the brain controls the right-hand side of the body and vice versa. The studies also demonstrated for the first time that the left and right hemispheres specialize in different tasks. For example, the left side of the brain is much more talkative than the right, which can produce only rudimentary words and phrases. It is more analytical and is better at math. The right-hand side deals with space, directions and music. It is much better at recognizing faces and allows you to understand the emotional content of language.

The work earned Sperry a Nobel Prize in 1981, and soon afterward a new theory of personality was born. It proposed that whichever side of your brain dominated determined whether you were logical and analytical, or creative and emotional. Even today, it is common to see the theory referred to in the popular press.

In fact, while the brain does indeed have discrete regions, each with a specific role, there is absolutely no evidence to suggest that any side dominates in the healthy brain. Think about language, for instance. Although the left side of the brain helps us to generate complex speech, the right-hand side gives it some finesse. Take the phrase "I'll show you the ropes." The left side of the brain is needed to produce the correct word order, but the right side of the brain provides you with the ability to understand and produce the verbal metaphor.

Rather than think about ourselves as being left- or right-brained, says Stephen Kosslyn, an emeritus professor at Harvard University, we should be thinking about ourselves in terms of our top and bottom brain, specifically how the two interact.

The top parts of the brain include most of the frontal cortex and the parietal lobes. The bottom parts involve some of the frontal cortex, but mostly the temporal and occipital lobes. When we split the brain up in this way, we can generalize about their roles, says Kosslyn. "The top brain formulates plans and puts them into action, while the bottom brain interprets incoming information about the world and gives it meaning."

It's vital to remember that we use both parts of the brain all of the time, says Kosslyn. "They are a single system—it's how they interact that's important."

For example, when I see my dad across the pub, I recognize him because the bottom parts of my brain interpret the sensory input I'm getting from my eyes, and give it context, which unlocks a memory of my dad. As we discovered in Bob's chapter, that memory is connected to other memories, which enables me to bring to mind the fact that he enjoys playing tennis, drinks Harvey's Best Bitter, and has a soft spot for Camilla Parker Bowles.

But that's not all I need from my dad. I might want to invite him to a quiz night, or ask his advice as an accountant. This is where my top brain speaks up. Devising and carrying out plans is its job—but it can't do this alone. It needs to receive information from the bottom brain about the thing I'm going to speak to him about and how I feel about it, in order to figure out a plan of action and then execute it. If that plan isn't quite working, my top brain will check in with the bottom brain again and adjust its actions to correct any mistakes.

The key to Kosslyn's theory is that, in some situations, we rely on the top or bottom to a greater or lesser degree. Which one dominates characterizes our personality.

For instance, if we thoroughly utilize both our top and bottom brain, we're inclined to carry out plans and think about the consequences in detail. But if the bottom brain system dominates, we're more likely to think about what we see around us in a lot of depth, interpreting the minutiae of an experience or the consequences of an event. Someone working in this mode is less likely to act on all this information and execute a plan. On the other hand, if your top brain system dominates you're more likely to be a go-getter, a person who may be seen as creative and proactive, but you're less likely to think about the consequences. Kosslyn calls it "the bull in the china shop" mode.

When neither top nor bottom brain dominates, a person doesn't get caught up in the detail of an experience nor initiate future plans. Instead they're very much "of the moment," says Kosslyn—they let external events dictate their actions. "They're a team player—not everyone can be president. You need people who are like soldiers, not looking beneath the surface of their actions, just getting on and doing what needs to be done at that precise moment."

If you want to find out which mode dominates in your brain, Kosslyn has devised a test you can take online at http://bit.do /topbrain.

In his book *Top Brain, Bottom Brain: Surprising Insights into How You Think*,[5] Kosslyn says that his theory can explain why we see sudden changes in personality. Take Phineas Gage as an example. As we discovered earlier, Gage suffered a terrible accident when a tamping iron shot through his skull. Historical notes suggest that before the accident Gage was unusually resourceful, that he was a good planner who learned from his experiences. It allowed him to work his way up the construction trade after barely any formal education. After his accident, there were profound changes in his personality. He would utter gross profanities and devise numerous plans that would immediately be abandoned for others.

After Gage passed away, his skull was donated to science and is now sitting in a small glass box at Harvard Medical School's Warren Anatomical Museum. Using the skull, researchers reconstructed the damage that occurred to his brain. They discovered that around 11 percent of the axons—the longest part of the nerve through which electrical impulses propagate—in his frontal cortex had been destroyed. This meant that a number of connections were severed between his frontal cortex high up in the brain and areas lower down.

It was not just that the trauma damaged specific behaviors, such as his ability to suppress verbal profanities, writes Kesslyn; the damage also impaired how his top and bottom brain worked together.

> Whereas Gage previously had been strategic and thoughtful, he now was impulsive and unstable. His bottom brain apparently interrupted his top brain inappropriately, impairing his ability to stick to plans or revise them after receiving new feedback on their effects. He was left awash in a sea of fluctuating emotions and was incapable of responding appropriately.

I wondered whether this theory could also explain what was going on in Tommy's brain.

TO FIND OUT, I decided to speak to somebody else who had developed a close relationship with Tommy after his accident—

Alice Flaherty, a neurologist at Massachusetts General Hospital in Boston.

Tommy had written to Flaherty soon after recovering from his operation to ask whether she could tell him anything more about his new personality. "He was such an attractive person," said Flaherty, when I asked her about their initial correspondence. "His letters were incredibly charming."

Flaherty tried to get Tommy over to her lab in the United States, but he couldn't get a visa due to his previous criminal convictions. In the end, she traveled to Liverpool for a few days. "I just loved him," she said. "He had this inability to hurt anyone, he was like a Jain monk who sweeps the ground ahead as he walks to avoid injuring bugs. He took care of every stray cat in the neighborhood. He used to have this tough-guy mask and suddenly he was a softie—in a lovely way."

Flaherty isn't keen on using Kosslyn's top-and-bottom-brain theories to hypothesize about what happened to Tommy. Instead she has made inferences based on what we know about where the areas of damage occurred. "In Tommy's case, we know that his stroke occurred in the middle cerebral artery, which supplies a portion of the frontal and temporal lobes," says Flaherty.

Although it is a matter of conjecture, the damage to his temporal lobes is the most likely explanation for his sudden obsession with the detail in the world. To understand why, we have to look at how the brain deals with the bewildering amount of sensory information around us. We see shapes, and colors and movement and hear sounds and smells all the time, yet rarely pay much attention to them. When I walk into the pub to meet my dad, I might notice the smell from the kitchen or the football on the TV at first, but within seconds these stimuli disappear. We filter out familiar and irrelevant information. If we didn't, our senses would be bombarded with too much information and we wouldn't be able to concentrate on the job at hand.

In order to do this, sensory data passes into the temporal lobes, which perform an emotional surveillance of sorts, telling other parts of the cerebral cortex whether the information is worth thinking about or not. Only the most relevant of data is sent up to the frontal lobes, which are then able to formulate plans, execute actions and initiate speech based on that information.

Tommy's behavior suggested that his brain had stopped filtering the irrelevant stimuli that are usually screened out of our conscious awareness. His temporal lobes were no longer judging all of his sensory data or ideas critically enough, says Flaherty, "so they all pass muster and float into his consciousness."

"We often see people with lesions in their temporal lobes losing their understanding of speech but talking loads more than normal," says Flaherty. "They've basically become less judgmental about what they're saying. We call it politician talk—lots of words but no content."

In contrast, Tommy's new emotional compass was more likely due to damage to the frontal lobes. These regions form connections with, and inhibit, emotional areas toward the lower and middle areas of the brain. The level of activity in the frontal lobes can affect our personality in many ways. In the 1960s, the German psychologist Hans Eysenck proposed that introverts may have more self-control than extroverts because their cortex is higher in arousal—meaning it is more sensitive and responsive to incoming information. This higher level of arousal keeps in check their emotional areas underneath.

You can test for yourself whether you are more introverted or extroverted. Place one end of a cotton swab on your tongue for twenty seconds. Next put a few drops of lemon onto your tongue before placing the other end of the cotton swab on your tongue for another twenty seconds. Tie a piece of thread around the middle of the cotton swab and see whether the lemon juice end is lower due to being heavier from excess saliva. If so, you may be more introverted—your higher arousal state meant that you reacted more strongly to the lemon, which made you salivate more than usual. Eysenck used a version of this test to support his theory by showing that people who score higher on other measures of introversion also produce more saliva.

A similar thing happens when you try to knock out an introvert with anesthetic—they need more to send them to sleep than an extrovert would. If you're still not convinced, think about how stimulants like Ritalin are given to children with ADHD to calm them down, and how sedatives like alcohol make people briefly more chatty and emotional.

Although it is a matter of conjecture, it seems that communication between Tommy's frontal lobes and lower brain areas had become impaired. Like Gage, damage to the frontal lobes seemed to have lifted the brake on his emotional brain areas below. Overnight, he gained access to emotions he said he "hadn't previously known existed."

Tommy's wife, Jan, thought that all these words, thoughts and emotions might be best placed on paper, so a few weeks after Tommy had been discharged from the hospital, she encouraged him to pick up a paintbrush. Maybe drawing what was in his mind might help him focus his thoughts. But once Tommy started painting, he couldn't stop.

"At first it was just lots of A4 pieces of paper taped to the wall," says Shillo. "We encouraged it because we all just thought it was part of his recovery. It was really helping." But soon Tommy ran out of canvases. First he bought more, but then that started getting too expensive, so he took to painting directly onto the walls of his house. Once he had covered all the walls in one room, he'd move on to the next. Once those were done, he'd move on to the floor, the tables and the chairs. Then he'd start all over again.

"We didn't live with him so didn't see it daily, but you'd go around and pretty much every month the house had completely changed, all the walls, the floor, everything," says Shillo. "The paint on the chimney breast was about two inches thick because he'd just paint layer upon layer of art over the top of each other."

"What's in my brain is painted on these walls," Tommy said. "The tables, the ceiling, the doors, in sculptures, in metal, in stone. Everything is poured out of my brain. I empty it on to my canvases with the color and pictures and scenes and I never stop."[6]

Tommy was painting for twenty-one hours a day. "We had to remind him to eat and sleep," Shillo says. "As long as he could paint and sculpt, nothing else seemed to matter much."

Tommy sent some pictures of his artwork to me. One was of two faces with images pouring out of them. "That painting was what he felt," says Shillo, when I describe it to her. "This drive, this desire, it was uncontainable, it was like looking at all these things that were firing in his brain that he couldn't control."

"Was he happy to see you, when you came around to visit?" I ask.

"Yes, he'd say that he missed us and it was lovely to see us and he'd be this wonderful person, but after a while you'd know the time had come to leave because he'd get fidgety, he'd want to get back to his painting. The moment you walked out of the door, it was like you didn't exist."

Tommy's endless creativity eventually drove Jan away.

"I don't blame her," said Tommy. "I was a completely different person."

"It was hard," says Shillo. "We all felt a little pushed out by his art, but we just accepted it—it was doing him so much good."

I asked Tommy whether his sudden love of art took him by surprise. Had he had any interest in it before?

"No, I'd never even picked up a paintbrush," he said. "Never been in an art gallery," he added as an afterthought. "Except maybe to steal something."

Flaherty knows personally what it is like to have a sudden and uncontrollable urge to do something creative. Following the death of her premature twins, she suffered from postpartum mania.[7]

"I couldn't sleep. All I wanted to do was talk, but because I'm an introvert I started writing it down instead," she told me. "Some primitive area of my brain was saying, 'Holy shit, something's going wrong, you gotta do something.'" She was flooded with ideas and for four months could do nothing but write. She realized that her mania resembled something called hypergraphia, which can occur alongside epilepsy and result in the intense desire to write. She decided to write a book about her experiences. "My writing a book was a way to understand myself," she said, "but also, if you're writing all the time and no one reads it then you're just crazy, but if you're an author then it's a good thing."

A year later, history repeated itself and she again gave birth to premature twins, though thankfully this time they survived. But once again she suddenly had the uncontrollable urge to write, mixed with periods of depression. Over the years, medicine and exercise have helped her control the condition.

The emergence of an uncontrollable creative drive has been termed sudden artistic output. It means that the brain can no longer inhibit certain behaviors. Tommy is one of a handful of famous cases in the scientific literature.

Another is Jon Sarkin, who in 1989 suffered a traumatic hemorrhage after an operation to fix a blood vessel that was pressing on his auditory nerve, making his ears ring. Months into his rehabilitation, Jon started to draw. Months turned into years and he became consumed by the need to paint all the time. He sold his chiropractor's office and became a full-time artist, whose work can sell for $10,000 apiece.

Then there's Tony Cicoria, an orthopedic surgeon who lives in upstate New York. In 1994, he was at a lake for a family gathering when he went to a pay phone to call his mum. Moments after he ended the call, a bolt of lightning struck him and he crashed to the ground. A nearby nurse brought him around with CPR. A month after the accident, Cicoria had returned to work, mostly feeling fit and well, when over a period of a few days he was overcome with a desire to listen to piano music. He began to teach himself how to play the piano and later began to hear music playing constantly in his head. Neurological scans at the time showed nothing amiss, and when offered newer techniques he politely declined, stating that his music was a blessing, and not something that he desired to question.

In 2013, I came across another case of sudden artistic output, in a paper that described a woman who had arrived at a hospital in the UK complaining of memory problems and a tendency to lose her way in familiar locations.[8] She was diagnosed with epilepsy and treated with the drug lamotrigine. As her seizures receded, a strange behavior took hold—she began compulsively to write poetry. She often used irregular rhythm and rhyme for comic effect, a style her husband described as "doggerel." I have one of her poems framed in my study:

To tidy out cupboards is morally wrong
I sing you this song, I tell you I'm right.
Each time that I've done it, thrown all out of sight,
I've regretted it.

Think of the treasures now lost to the world
Measureless gold, riches unfurled,
Diamonds, sapphires, rubies, emeralds—you must have
    had them,
All tucked well away. So

To tidy out cupboards, throw rubbish from sight
(Even the poems you write up at night)
Is morally wrong.
So I'm keeping this one.

The relationship between art and brain damage is a complex one. The reasons suggested for these rare cases of sudden artistic output are speculative at best, but Flaherty proposes that they may include an increase in dopamine. Dopamine is used all over the brain and is important for motivation and producing a drive toward things that make you happy. If you have too much dopamine, though, it can lead to disinhibited behavior. Those affected might start gambling, partake in riskier activities and show compulsive behaviors, including a sudden urge to draw or play music. It is a side effect that has been noted in people with Parkinson's who are on high doses of L-Dopa, a drug that increases dopamine to replace that lost through the disease.

Disinhibition of brain pathways is also seen in Tourette's syndrome, whereby people find it difficult to suppress inappropriate words and noises. Interestingly, this behavior is often accompanied by a substantial creative drive.

It's not possible to say exactly what Tommy's trigger was, says

Flaherty. Perhaps the stream of ideas and concepts was his muse; perhaps an increase in dopamine created the obsession, perhaps putting paintbrush to paper helped him make sense of it all. Whatever the cause, she says, "It was clear that painting gave him an extreme sense of well-being."

I often wondered whether Tommy's new personality allowed him to reflect on how he used to be, how he had treated his family. But despite having told me about several aspects of his past—his childhood, the drugs and the fights—when I asked him these questions, he said he remembered almost nothing about his old life.

"I sometimes hear my mammy's voice," he said, "and suddenly get a taste of a past, but it's not really remembering. People have told me about my life by their stories, but I don't know what the hell they're talking about."

"You don't remember how you used to act?" I asked.

"No. When I woke up I found it difficult to recognize people. Bits about my childhood came back but not everything. I found out a lot through stories that people tell me. But sometimes I think they've made the stories into a kind of Chinese whispers. Their stories become bigger, so I take no notice of them. My memory starts at 2001."

But when I ask Shillo about this apparent memory loss, she tells a different story. "Dad would often apologize for how he had been in the past," she says. "He would always say his memory wasn't that great, but in reality, if you were reminiscing about the past, his memory was incredibly good.

"I think in reality, he just didn't want to spend too much time delving into his memory to see what was there. He didn't want to fully remember who he was before, because he was so emotional after the accident, the things he'd done would have been a lot harder to deal with."

I ASKED TOMMY whether he preferred his new personality to what he remembers of the old one.

"The best thing that happened to Tommy McHugh," he said, "was having a stroke while doing a poo."

I laughed.

"You just have to accept this alien, unknown identity," he said. "Then you adapt and start to live again."

Tommy said he felt that during his rehabilitation a lot of his doctors were trying to find the old Tommy, rather than embrace the new one. "Helen, none of this is just about me," he said. "There's a world of people like me living out these strange new worlds of brain repair. And a lot of them don't have anything or anybody to help them express what's going on."

Tommy was so impassioned by this issue that he started giving talks to other stroke survivors, to encourage them to embrace their new mind, rather than trying to capture all the aspects of their old one.

"Our brain is repairing itself, and sometimes that's in new, constructive ways and sometimes it's in negative ways. We need to be able to talk about the strange things that happen to us, because getting a bit of support and understanding in recovery can make 100 percent difference. What a lot of people forget is that we're alive, we've survived this really hard adventure.

"Those of us who have had strokes and can walk and talk need to let other people know that it's not the end of the world, it's a beginning—a chance to improve one's mind and not be put on a shelf and labeled as a brain-damaged buffoon."

I ASK SHILLO whether she thinks her dad seemed happier after the stroke.

"Definitely," she says. "His old behavior was a result of this heaviness that he felt. There used to be this switch in his head that you'd see when he would suddenly realize he had gone too far and he'd think"—she whispers as Issac is in earshot—"'I fucked up, I might as well go the whole way and take everybody down with me.'

"He would ruin everything because he thought he'd already gone too far into this lonely, dark and scary place. After his strokes, he was certainly a much more balanced, settled, happier person. He liked himself, whereas I don't think he did before. No matter how frustrated he got, he never got dark, he never let it send him down. He'd just step back and take another route. It took us a long time to accept that there wasn't going to be any more dark times."

Redemption, that's what it came down to, says Shillo suddenly, as the rain begins to fall heavily against the kitchen window and I notice how late it has gotten. "It was a chance for him to make things right." To make up somehow, perhaps unconsciously, for the "dark times" that he had once led.

"People think a horrific brain injury is the end and I'm not so sure it was for Dad. It gave him a fresh start and there's not many people who are capable of having that in life. He was able to wipe the slate clean, and he took advantage of that. He started again as a good person."

It's not often that we take the time to consider our personality, who we are and how we make our choices. Perhaps it's because we tend to think our personalities are innate, that they are what they are. I can't help wondering whether knowing more about the mechanisms that help to build them could help us navigate life a little more successfully. Perhaps even make us all a little happier.

In 2007, a team lead by Anita Woolley answered part of this question. She started by giving almost 2,500 people a questionnaire that assessed their ability to think about the properties of objects or their spatial location—a method that allowed her to figure out whether they were "top-brainers" or "bottom-brainers." For example, participants were asked to recount what someone had been wearing at a recent dinner party—an answer that is accomplished primarily using the bottom brain, which stores visual memories of color and shapes. They were also asked questions that involved spatial manipulation, such as imagining how the Statue of Liberty might look if rotated. This spatial imagery is accomplished primarily by the top brain. Woolley's team then chose 200 people who scored highly in the top-brain tasks but low in the bottom-brain tasks, or vice versa.

Next, they split the group into pairs and asked them to navigate a virtual maze. At various sections of the maze were little greebles. Greebles are computer-generated objects that don't obviously resemble anything in real life. Sometimes a particular greeble would pop up twice in different parts of the maze.

One of the pair had to navigate the maze using a joystick; the other was tasked with tagging duplicated greebles of the same shape. Unbeknown to the volunteers, each was given a role that suited their strengths or deliberately went against them. So in some trials a person who was great at using their spatial top brain would be asked to navigate, and their partner—a bottom-brainer who was good at object recognition—would tag greebles. In the opposite condition their roles were reversed. In a final set of experiments, both people in the pair were top- or bottom-brainers.

As you might expect, when the roles were suited to their brains the teams performed best. But here's the rub: this occurred only when the task was completed in silence. In trials where the pairs were allowed to talk to each other, the incompatible teams suddenly did as well as those who were naturally suited to their tasks.[9] When the researchers watched back the trials, they discovered that each person quickly took over the other's role, helping them to achieve their assignment. In other words, complete strangers had spontaneously discovered their strengths and weaknesses and modified their behavior to get a job done.

Funnily enough, when two people who were both top-brainers or both bottom-brainers were allowed to talk, their results suffered more than when they worked in silence. Two people with the same abilities, trying to help each other with something neither was good at, just made matters worse.

What this experiment suggests is that it's useful to know what kind of personality traits we possess, whether it's using the Big Five or Kosslyn's top-brain and bottom-brain modes. With that knowledge, we can be more productive both at work and at home. While none of us would want to experience a personality change as dramatic as Tommy's, we may want to tweak our personalities every now and then to suit a particular challenge better.

For instance, if you come across a problem that you can't solve, try thinking about it using the kind of strategy that doesn't come easily to you. "It takes more effort to think in a different mode," says Kosslyn, "but anyone can drop into any mode if you really try."

Or, like the people in Woolley's experiment, you could augment your skills and knowledge by working with people who have the skills that you lack. It's as if you've borrowed a part of your companion's brain, says Kosslyn, and by doing so have extended your own reach and capacities.

When I returned home from Shillo's, I took Kosslyn's online test to see what part of my brain dictated my life. It suggested that I have a strong tendency to rely on my top brain, and a lesser tendency to rely on my bottom brain. It rang true—I thoroughly enjoy making plans and carrying them out, but I am often too quick to pass over the finer details. My husband is the complete opposite, he is strongly bottom-brain biased—he's very good at thinking about the details, but shies away from using them to initiate any plans. That particular trait used to frustrate me. But now I'm starting to see our relative skills and deficiencies in a different light. Together our personalities make the perfect team.

In September 2012, a few months after our final correspondence, and more than a year before I would come to meet his daughter, Tommy died of liver disease. When I heard of his death, I reread all of our conversations, emails and letters. The last email I ever received seemed a good place to start:

You have a new message from Tommymchugh2:

I look at my reflection in the mirror Helen. A stranger I see. But a happy one, xxxx to all.

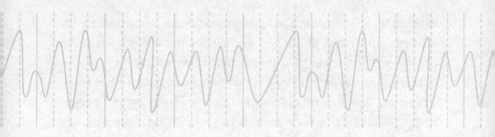

SYLVIA

An Endless Hallucination

Avinash Aujayeb was alone, trekking across a vast white glacier in the Karakoram, a mountain range on the edge of the Himalayan plateau known as the roof of the world. Earlier that morning he had left his two companions after feeling too exhausted to finish the journey to the summit. Faced with no option but to head back to camp, he set off down the mountain. He had been walking for hours, yet his silent surroundings gave little hint that he was making progress.

Suddenly, everything changed. A gigantic icy boulder loomed close one moment then far away the next. He glanced from left to right but couldn't shake the feeling that he was looking at the world over his own shoulder. Concentrating on placing one foot in front of the other, he set himself small goals of reaching the next ridge or rocky outcrop. It seemed to take an hour to get from one checkpoint to the next, yet when he looked at his watch, only minutes had passed.

As a doctor, Aujayeb made a mental checklist. He wasn't

dehydrated, nor did he have altitude sickness. He checked his heart rate and blood pressure—they were completely normal. So why couldn't he shake the notion that he was dead?

Aujayeb was experiencing a vivid and prolonged hallucination, something that before 1838 was simply referred to as "a wandering mind." It was the French psychiatrist Jean-Étienne Esquirol who first characterized a hallucination when he described it as something experienced by someone who "has a thorough conviction of the perception of a sensation, when a non-external object, suited to excite this sensation, has impressed upon his senses."[1] Or in other words, who saw something that wasn't really there.

Hallucinations aren't limited to images; they can appear as music, voices and even smells. They can last for seconds or for months, and are likely to have shaped our culture, religion and societies for centuries. In his book *Hallucinations*, Oliver Sacks wonders whether Lilliputian hallucinations, in which objects, people or animals seem smaller than they should be in real life, may have given rise to the elves, imps and leprechauns in our folklore. He suggests that terrifying hallucinations of a malign presence may have sparked the idea of demons, and that out-of-body hallucinations or hearing voices may have generated our sense of the divine.[2]

In the past, particularly in Western cultures, there has been a tendency to view hallucinations as a sign of mental malfunction. Yet in recent years episodes like those experienced by Aujayeb have forced scientists to rethink their ideas of hallucinations as purely a symptom of mental illness or the result of mind-altering drugs. They have begun to realize that hallucinations are neither uncommon nor always a sign of ill health.

One person who knows a great deal about this is Sylvia, a retired math teacher from north London. Despite being incredibly sharp, with excellent mental health, Sylvia has experienced a permanent hallucination every day of her life for the past decade. One morning in midwinter I went to meet her to find out more about this strange phenomenon. In learning more about her life, I came across the most startling discovery of my journey so far. I found that hallucinations are not only common, but vital to producing our perception of reality. So vital, in fact, that you're probably hallucinating right now.[3]

It's difficult to imagine what a hallucination feels like unless you have experienced one yourself. I say this with some degree of conviction because a few months ago, early in the morning and alone in bed, I was woken by two strangers entering my room.

I was paralyzed with fear, I felt wide awake, yet I couldn't move my body. One of the strangers, a man, walked to the opposite side of the room, while the other, a woman, sat down at the bottom of my bed. As she did so, I felt the covers brush across my legs. Later, I discovered that I had been experiencing what's known as a hypnopompic hallucination. These appear during the transitory period between a sleeping state and wakefulness, and probably occur due to certain parts of the brain remaining in REM sleep—the period in which we dream the most—while other areas become fully conscious. For me, it had a distinct feeling of objective reality. It was far closer to the experience of someone actually being in the room than the experience of dreaming they were there.

This distinction is supported by some evidence. In 1998, Dominic Ffytche, a senior lecturer in old-age psychiatry, and his colleagues at King's College London scanned the brains of people experiencing visual hallucinations. They found that the areas of the brain that were active are also active while viewing a real version of the hallucinated image. Those who hallucinated faces, for example, activated areas of the fusiform gyrus, known to contain specialized face-recognition cells active when we look at real faces. The same was true with hallucinations of color and written words. When the team asked the participants to imagine faces, or colors and words, there was nowhere near as much activity in the corresponding areas of the brain. It was the first objective evidence that hallucinations were less like imagination and more like real perception.[4]

As well as hynopompic hallucination, like the one I experienced, another common hallucination is seeing shapes or hearing sounds as you drift off to sleep at night. Or seeing visions of loved ones during a period of grieving. But the kind of hallucination that I am most intrigued about—and the one that reveals most about the brain—is that which crops up in people who have recently lost a sense.

Several years ago, my mum called me to tell me that my grandmother had started seeing people. Eighty-seven years old, she began to hallucinate after her already poor sight got worse due to cataracts. Her first visitors were women in Victorian dress, shortly followed by young children who danced around her bedroom. Sometimes all she saw was a plain brick wall. My nan was not, it seemed, bothered by these hallucinations; she knew they weren't real, despite their vividness, but was intrigued by what they meant.

She was experiencing a common condition in those with failing sight called Charles Bonnet syndrome. Bonnet, a Swiss scientist born in 1720, became interested in hallucinations when his grandfather started to experience visions soon after his eyesight began to fail. One day, his grandfather was sitting in an armchair talking to his two granddaughters. Suddenly two men appeared. They were, by all accounts, wearing magnificent cloaks of red and gray with hats trimmed with silver. When he asked his grandchildren why no one had told him they would be coming, the elder Bonnet discovered that only he could see them.

Over the next month, his visions appeared on many occasions. Sometimes they were further beautiful visitors, other times they were pigeons or butterflies, sometimes simply oversized carriages. Bonnet's grandfather apparently enjoyed what he referred to as "the theater in his mind,"[5] watching these apparitions for several months before they disappeared completely. Bonnet later experienced the same condition when he too lost his sight in old age.

It was a similar story for Max, a man I interviewed for the BBC in 2014. Max was in his seventies when Parkinson's disease destroyed the nerves that send information from the nose to the brain. Despite his loss of scent, one day he suddenly noticed the smell of burning leaves. He was on holiday at the time and so glanced around his hotel room, wondering what could have caused the strange aroma. He was convinced there was a family of skunks around.

"It was so strong," he said. "It gave me this strange sensation in my throat that I couldn't get away from."

Over the next few weeks, the aromas intensified, ranging from burnt wood to a horrible onion-like stench. They stuck around after he returned home too, sometimes persisting for hours.

"When they're at their most intense they can smell like excrement. It's so powerful it makes my eyes water."

Sensory loss doesn't have to be permanent to bring on such hallucinations. Avinash, after all, was in fine health while trekking across the glacier.

"I knew I wasn't sick," he told me. "My heart rate was fine. I wasn't dehydrated and I'd had enough food. I tried to rationalize it all. I pinched myself, I kept trying to make sure I wasn't asleep or dreaming. At one point I tripped and cut my hand— the sight of blood made me sure it wasn't a dream."

At some point he started hearing a voice, and felt like there was a presence guiding his every move. "I felt as if it was talking me through things. It was asking me to think carefully, to pick my way through each glacier. It was helping me, guiding me toward where I should be."

The whole experience lasted almost nine hours.

"I asked myself at one point, 'Am I dead?' This was a difficult trek—you could have just fallen into a crevice and died and no one would ever have found you. It was only when I passed someone else on the mountain that I knew I was definitely alive. But even when I met up with the rest of my group I still felt strange. It was only after a good night's sleep that everything went back to normal."

Searching for answers to his weird hallucinations, Avinash momentarily wondered whether he had achieved the "savikalpa samadhi"— a state said to be reached by meditation in Buddhist or Hindu tradition through which one is said to lose all human consciousness, and in which the conception of time and space is altered.

It turns out the answer is far simpler. But to understand more, I needed to speak to Sylvia.

One Friday morning in 2004, the people of Potters Bar were going about their daily business. Minutes from the center of town was sixty-year-old Sylvia, a retired teacher, who was working in her house. All was well, apart from one thing. That wretched noise. The two notes that had started playing earlier that morning, which no one else seemed to be able to hear. Sylvia had initially thought the radio must be playing, but a quick search proved that not to be the case. Fairly fearful of the strange new noise, which got louder throughout the day, Sylvia went to bed, hoping it would be gone by morning. She woke up to find the noise was still there. It droned on and on, *dah de dah de dah de.* Over weeks, the notes changed, and months later finally became full-blown musical hallucinations—tunes that were constantly in the background, sometimes so loud that they drowned out normal conversation.

"Please ignore her for a moment," Sylvia says, as she welcomes me into the house. She's talking about the golden Labrador sitting quietly in the hall. Sookie is Sylvia's new hearing dog.

"Good girl," she says to the dog. "Now you can go say hello." Sookie bounds over to me and sticks her nose straight into my pocket. "She thinks you may have treats," says Sylvia. "It's always worth a try."

Sylvia needs a hearing dog because she is deaf. She finds it difficult to hear speech, and real music sounds awful, distorted by profound hearing loss caused by an ear infection several years earlier.

Sylvia's husband, John, waves hello as we walk past a grand piano and through to a bright conservatory at the back of the house. I sit down on a wicker chair, while Sylvia serves up tea and biscuits.

She takes me back to the Friday morning that it all started. She'd had tinnitus and hissing for a number of years previously. But this was something new. It was the notes C and D going back and forth, she says. "It was very slow at first. I remember thinking, 'Oh I don't want that, think about something else.' The notes grew from there. I've never had quiet since."

Gradually over weeks the notes developed into short phrases that would repeat over and over again. Sometimes they would get longer, forming complete melodies from the music she had loved before she lost her hearing.

"What kinds of tunes do you hear most often?" I ask.

"It's mostly classical music, short excerpts. When I could hear properly, I didn't tend to listen to much else."

Even as we sit talking—through the help of a microphone and lip reading—Sylvia's tunes are playing in her head. If they ever recede, which they sometimes do when she's focused on a piece of music or concentrating on speaking, they are replaced by a constant B-flat tone and the hiss of tinnitus.

"Do they resemble any particular instrument?" I ask.

"They're a cross between a wooden flute and a bell," she says. "It's really strange—you'd expect to hear a sound that you recognize, maybe a piano or a trumpet, but it's not like anything I know in real life."

"But it sounds like a real noise?"

"Yes, it's not as if you imagine a tune in your head, it's like listening to the radio. It manifests as real sound."

Soon after her hallucinations started, Sylvia did something very constructive. She wrote them down, keeping a record in a manuscript book of all the tunes that would come and go. She has the rare talent of perfect pitch, which allows her to hear any note and know exactly which note is being played.

She brings the sheet music to the conservatory for me to read. Some of the pieces are made up of random notes, never going anywhere in particular. Other hallucinations resemble short sections of recognizable tunes—I spot a passage from the traditional Scottish folksong "My Bonnie Lies Over the Ocean."

Seeing her hallucinations on paper highlights their repetitiveness. There are pages of notes that wind up and down, up and down. They are like this for most of the day, Sylvia says.

Years of teaching math allows her to make a quick mental calculation. "If it's just two or three notes then they can take a second to play before they repeat," she says, "so that's what?

Hearing the same tiny tune about eighty-six thousand times a day?"

Sylvia tells me that early on in the evolution of her hallucinations, words began to put themselves to the tunes.

"I did my utmost to prevent that happening," she says, "and I managed to make it stop."

I ask her why.

"I didn't want that happening. I didn't want words coming into my mind. It just felt a bit close to schizophrenia."

SHE'S RIGHT, OF COURSE—hearing voices that are not there is often considered a sign of mental illness. No one knows this better than David Rosenhan, a professor emeritus at Stanford, who in 1973 got himself and seven other completely healthy friends admitted to the mental wards of hospitals across the United States. The point of his experiment was to question the validity of psychiatric diagnosis—but they hadn't expected it to be so easy. Rosenhan and his colleagues each phoned a hospital complaining of hearing voices. The rest of their medical history and any other life stories were all true. All eight were admitted—seven were diagnosed with schizophrenia, one with manic-depressive psychosis. As soon as they entered the hospital they said their hallucinations had disappeared. It was then up to each individual to convince the staff to discharge them—a task that took between seven and fifty-two days.[6]

In fact, most hallucinations are not associated with schizophrenia. When John McGrath, a professor at the Queensland Brain Institute in Australia, analyzed interviews with more than 31,000 people from eighteen different countries, he discovered that hallucinations were fairly common in all age groups. When participants were asked whether they had ever experienced a hallucination, such as hearing voices that other people said did not exist, 5 percent of men and 6.6 percent of women answered yes.[7]

I ask Sylvia whether she tells other people about her hallucinations.

"No, I don't talk to many people about it. Very early on, I was taught that as sound travels up to the brain it picks up emotions that go with that sound. In other words, if I became constantly irritated by them, the sounds would always cause irritation. If I told them they were unimportant, they would stay insignificant. So I took a conscious decision to belittle them, and not talk to other people about them. I don't want them to garner an importance. It was the best advice I could have been given—it's allowed me to live with them."

She smiles. "Sometimes I go, 'Oh shut up,' and my friends know what I'm talking about, but they can't possibly put themselves in my shoes."

She glances at John as he pops his head into the conservatory. "John, bless him, is a real support in every sort of way, but even he has no idea how dramatic this sound is that's going on all the time. It gets in the way of his conversation with me. I mishear what he says so much of the time. Sometimes I think he's said something very funny and he didn't say it at all. He's very understanding, but no one can have any idea what this is like unless they've experienced it for themselves."

In fact, there is a way that you can experience something similar for yourself, safely at home. All you need is a table-tennis ball, some headphones and a bit of tape. Cut the ball in half and tape each segment over your eyes. Sit in a room that is evenly lit, find some white noise to listen to over your headphones, sit back and relax.

Called the ganzfeld technique, this kind of sensory deprivation has been used to investigate the appearance of hallucinations for decades. In a paper published in the journal *Cortex*,[8] Jiří Wackermann at the Institute for Frontier Areas of Psychology and Mental Health in Freiburg, Germany, describes some of the hallucinations that occurred in volunteers trying out the technique.

"For quite a long time there was nothing except a green-greyish fog," said one participant. "It was really boring. I thought, 'Ah, what a nonsense experiment!' Then, for an indefinite period of time, I was 'off,' like completely absent-minded. Then, all of a sudden, I saw a hand holding a piece of chalk and writing on a blackboard something like a mathematical formula. The vision was very clear, but it stayed only for a few seconds and disappeared again . . . it was like a window into that foggy stuff." Later, she saw a clearing in a forest and a woman who passed by on a bike, her long blond hair waving in the wind.

Another participant felt like she and a friend were inside a cave. "We made a fire. There was a creek flowing under our feet, and we were on a stone. She had fallen into the creek, and she had to wait to have her things dried. Then she said to me: 'Hey, move on, we should go now.'"

As I sit back in my own living room, ping-pong balls strapped to my face, I feel much the same way as that first participant. Nothing happened for at least thirty minutes, other than a myriad of random thoughts and waves of sleep. Just as I was wondering whether I should give up, I saw an image coming out from what seemed like a window full of smoke. It was of a man lying curled up next to me. He bent his elbow in this strange way as if presenting it to me. It appeared for a few seconds, then disappeared. It certainly differed from a dream, or from a random image plucked from my imagination. It was an intriguing demonstration of what can occur when our senses are impaired. But why does it happen?

"The brain doesn't tolerate inactivity," said Oliver Sacks, when I spoke to him about this back in 2014. "It seems to respond to diminished sensory input by creating autonomous sensations of its own choosing."

This was noted soon after the Second World War, he said, when it was discovered that high-flying aviators in featureless skies and truck drivers on long, empty roads were prone to hallucinations.

Now researchers believe that these unreal experiences provide a glimpse into the way our brains stitch together our perception of reality.

Although bombarded with thousands of sensations every second of the day, your brain rarely stops providing you with a steady stream of consciousness. Think about all the sounds, smells and tactile sensations that you can sense right now. Noise from the outside world, the tightness of your socks, the feel of this book against your fingers. Processing everything that you experience in the world all of the time would be a very inefficient way to run a brain. So instead it takes a few shortcuts.

Let's use sound as an example. When sound waves enter the ear, they are converted into electrical messages by receptors within the ear canal and transmitted to the brain's primary auditory cortex. This part of the brain processes the rawest elements of sound, such as patterns and pitch. From here, the signals get passed on to higher brain areas that process more complex features, such as melody, key changes and emotional context.

Instead of relaying every detail up the chain of command, the brain combines the noisy signals coming in with prior experiences, to generate a prediction of what's happening in the world.

If you hear the opening notes to a familiar tune, for instance, you expect the rest of the song to follow. That prediction is passed back to the lower regions, where it is compared to the actual input, and to the frontal lobes, which perform a kind of reality check, before it pops up into our consciousness. Only if a prediction is wrong does a signal get passed back to higher areas, which adjust subsequent predictions.

You can test this for yourself. Anil Seth, a cognitive and computational neuroscientist at the University of Sussex, suggests listening to sine-wave speech, a degraded version of a speech recording. The first time all you'll hear is a jumble of beeps and whistles. But if you listen to the original recording and then switch back to the degraded version, you will suddenly be able to make out what is being said. All that has changed is your brain's expectations of the input. It means it now has better information on which to base its prediction. "Our reality," Seth once told me, "is merely a controlled hallucination, reined in by our senses." Or, as psychologist Chris Frith once put it, "A fantasy that coincides with a reality."[9]

THIS IDEA IS CONSISTENT with what was happening to Sylvia.

Although her regular hearing is distorted, familiar music can sometimes suppress her hallucinations for brief periods. In 2014, Timothy Griffiths thought he might be able to use this to support the prediction model of hallucinations.[10]

"The main obstacle to studying hallucinations and why they arise has always been an inability to control them—Sylvia gave us a chance to switch them on and off," he says.

Griffiths and his colleagues had Sylvia come into their lab and lie down in a machine that analyzed her brain waves—the cyclic flow of electrical activity around the brain. While the machine was analyzing her brain activity, Griffiths's team played

Sylvia various passages from a familiar Bach concerto. Sylvia rated the intensity of her hallucinations every fifteen seconds throughout the study. At the time of the experiment, her musical hallucinations happened to consist of sequences from Gilbert and Sullivan's musical *HMS Pinafore*. Immediately following the Bach, her hallucinations were silent for a few seconds, gradually increasing in volume until the start of the next excerpt. It was this that allowed Griffiths to measure her brain activity without the hallucinations and then with them.

Sylvia's brain scans showed that, during her hallucinations, regions of the brain that process melodies and sequences of tones were talking to one another just as they might if she was listening to real music. Yet, because Sylvia is severely deaf, they were not constrained by the real sounds entering her ears. Her hallucinations are her brain's best guess at what is out there.

This theory also explains why listening to some kinds of music can stop Sylvia's hallucinations. When she is concentrating on Bach, something she is familiar with, the signal entering her brain is much more reliable and that constrains the aberrant conversation going on in higher areas; this then reconciles itself to what is actually happening in the real world.

This notion of hallucinations as errant predictions has also been put to the test in completely silent rooms known as anechoic chambers. At Orfield Laboratories in Minneapolis, Minnesota, you can find such a chamber, dubbed "the quietist place on earth." The chamber is actually a pitch-black room built inside a room, built inside another room. It has three-foot-thick walls of steel and concrete and is lined with jagged padding, designed to absorb what's left of any tiny sound. Once inside, you can hear your eyeballs moving and your skin stretching across your skull. People generally start to hallucinate within twenty minutes of the door closing.[11] But what's the trigger?

I asked Oliver Mason, a clinical psychologist at University

College London who specializes in sensory deprivation. There are two possibilities, he said. One is that sensory regions of the brain occasionally show spontaneous activity that is usually suppressed and corrected by real sensory data coming in from the world. In the deathly silence of an anechoic chamber, under the influence of the ganzfeld technique or in the case of a permanent lost sense, the brain may make predictions based on this spontaneous activity that then run riot. The second possibility is that the brain misinterprets internally generated sounds. In an anechoic chamber, for instance, the sound of blood flowing through your ears isn't familiar, so it could be misattributed as coming from outside you. "Once a sound is given significance, you've got a seed," said Mason, "a starting point on which a hallucination can be built."

NOT EVERYONE REACTS in the same way inside an anechoic chamber. Some people don't hallucinate at all. Others do, but realize it is their mind playing tricks.

"Some people come out and say, 'I'm convinced you were playing noises in there,'" says Mason.

This was something I was puzzled about—why did Sylvia hear hallucinations when others with hearing loss did not?

When I asked Mason about this, he told me there were several theories. Finding out the answer is incredibly important, he said, because it could reveal why some people are more prone to the delusions and hallucinations associated with mental illness.

We know that electrical messages passed across the brain are either excitatory or inhibitory—meaning they either promote or impede activity in neighboring neurons. In a recent unpublished experiment, Mason's team analyzed the brain activity of volunteers while they sat in an anechoic chamber for twenty-five minutes. Those who had more hallucinatory sensations had lower levels of inhibitory activity across their brain. Perhaps, says

Mason, weaker inhibition makes it more likely that irrelevant signals suddenly become meaningful.

People with schizophrenia often have overactivity in their sensory cortices, but poor connectivity from these areas to their frontal lobes. This might mean that the brain makes lots of predictions that are not given a reality check before they pass into conscious awareness, says Flavie Waters, a clinical neuroscientist at the University of Western Australia in Perth. In conditions like Charles Bonnet syndrome, it is underactivity in the sensory cortices that triggers the brain to start filling in the gaps, and there is no actual sensory input to help it correct course. In both cases, says Waters, the brain starts listening in on itself, instead of tuning into the outside world.

This kind of research is helping people like Max, who can spend whole days surrounded by strange smells, reconnect with the external world. If his smell hallucinations are driven by a lack of reliable information, then real smells should help him to suppress the hallucinations. He has been trialing sniffing three different scents three times a day. "Maybe it's just wishful thinking," he says, "but it seems to be helping."

The knowledge that hallucinations can be a by-product of how we construct reality might change how we experience them. In his later years, Sacks himself became blind in one eye and had severe vision loss in the other. When he played the piano, he noticed that he would occasionally see showers of flats when he was looking carefully at musical scores. "Why they are flats, not sharps, I don't know," he said. He also hallucinated letters and the occasional words.

His hallucinations didn't bother him, he told me. "I have long since learned to ignore them, and occasionally enjoy them. I like seeing what my brain is up to when it is at play."

Recently, Sylvia's musical hallucinations have become much quicker, so that the notes go at a faster pace. They've also become louder. Now, she says, her hallucinations have developed so much that if she practices a Mozart sonata on the piano and then stops, the whole of the first movement will play in her mind. She says it's like having one's own internal iPod. It has its downsides—December can become a nightmare for Sylvia: "There'll be carols in all the supermarkets so I'll get fragments of those playing over and over—it will drive me mad."

Interestingly, words have also started to influence her hallucinations. The previous day, Sylvia had been reading, and the word "abide" was on the page. Suddenly the hymn "Abide with Me" had started to play on Sylvia's internal iPod. Images can also trigger songs. When she was in a toy shop with her granddaughter, she caught sight of a jester with bells hanging off its hat. Suddenly, "When That I Was and a Little Tiny Boy" started playing—the jester song from *Twelfth Night*.

She says she has also begun to get a modicum of control over her hallucinations. That morning, for example, she had gone swimming. The wax earplugs make everything silent, which renders the tunes in her head even more obvious. "It was going 'yada da bomb bomb, yada da bomb bomb,'" she says. "I didn't want that going on all the time I was swimming so I pitched out loud a note that was a semi-tone up from what was playing to conflict with it. This made the tune hesitate. It sometimes takes a long time, but I can often make it change. I can also change it by singing another tune that I'd prefer to hear. Sometimes it works, sometimes it won't. Sometimes it changes for a bit and then reverts back to the original annoying few notes—it's like a stubborn child in there, saying, 'No, I want to play this.'"

I ask her if she ever has more than a few seconds of silence.

"No, never," she replies.

"Do you ever feel like you can tune into your own private radio and enjoy what you're hearing when it's a tune that you like?"

Sylvia thinks about this for a while. "I've been terribly careful not to let the tunes pick up any emotion, so that they don't constantly make me emotional," she says. "I mean they do still make me irritable. Sometimes I feel as if I haven't fully rested when I wake up, sometimes it's really intrusive when they start playing before you've even found your slippers. But perhaps that's just me being an irritable old woman! But I don't mind it so much when it's a full tune that I recognize." She smiles. "I laugh at it a bit. I listen to it. I marvel at it. I do my best not to sing to it in case it reinforces it."

She pauses. "But then it contracts. It always contracts. It might play two or three times and then it gets shorter and I realize it's only the first two pages, or the first two lines of the tune and eventually just two or three notes again. That's the point when you can truly imagine being driven mad by it. Just 'dah di dah *dah* dah, dah di dah *dah* dah, dah di dah *dah* dah, dah di . . .'"

I leave Sylvia's later that afternoon amazed by her control, resilience and good humor over what could easily have become a soul-destroying condition. Society teaches us to be fearful of those things that are not present in the world, to associate seeing and hearing things that no one else does as a sign of mental unrest. Sylvia, Avinash, Max—even my own nan—prove that this need not be true. We shouldn't be afraid to fight this misunderstanding, to speak out when we experience things a little unusual. It is possible that we are all hallucinating all of the time—some of us are just more aware of it than others.

MATAR

Turning into a Tiger

Throughout history there have been legends of men who could turn into animals and then back into human form. The most feared of all is the werewolf, a bloodthirsty creature beset by murderous urges, devouring both the living and the dead.

This man-to-beast story has appeared in almost every period of human history—from our earliest popular fiction, *Satyricon*, to the Roman tales of Lycaon, the cruel leader of Arcadia, who was transformed into a wolf as punishment for trying to trick Zeus, god of the sky. Today, we only have to turn the pages of Harry Potter or the Twilight saga to see that the werewolf's tale has lost none of its gory appeal.

You may wonder where werewolves fit into my quest to meet people with the world's strangest brains. But the extraordinary truth of it is, werewolves aren't restricted to popular fiction and folklore—there are references to people turning into animals in some of our earliest medical texts. Paulus Aegineta, an Alexandrian physician in the seventh century, described the affliction as something suffered by people with melancholy or an excess of black bile. Increasingly over the medieval period, it was interpreted as the work of magic and the devil. The result was a person who was said to be prone to beast-like howls, who would seek out raw meat and attack other humans.

What could have caused such an affliction? One possibility is that ointments prescribed at the time for other illnesses could have led to side effects akin to chronic pins and needles. This may have been interpreted as the feeling of hair growing inside the skin and "proof" of a person turning into an animal.

Historians have also suggested that ingestion of medicinal plants, such as poppies or henbane—a plant similar to toxic belladonna—might have been to blame. Seventeenth-century herbalists used henbane as a sedative, and as a cure for rheumatic pain and toothache. We now know that these treatments can produce vivid hallucinations. There are extensive accounts of people feeling like they've been temporarily transformed into leopards, snakes and mythological animals after ingesting such plants.

Over time, several cures were considered, which included drinking vinegar, purging the body of blood and, most drastic of all, being shot with a silver bullet.

One of the most famous werewolf accounts is that of fourteen-year-old Jean Grenier, from Les Landes, France. In the early seventeenth century, Grenier boasted of having eaten more than fifty children. He said he preferred to run around on all fours and felt cravings for raw flesh, "especially for that of little girls," which, he claimed "is delicious."[1] Grenier was sentenced to be hanged and his body burned. However, before this could happen, the local council sent two doctors to examine him. They decided he was suffering from "a malady called lycanthropy—induced by an evil spirit, which deceived men's eye into imagining such things."[2] Rather than face execution, Grenier was sent to a monastery.

It wasn't until the mid-nineteenth century that a completely rational explanation prevailed, with physicians concluding that the condition was not mystical in nature, but a form of mental illness. In the past century, what is now known as clinical lycanthropy has been given a broader definition, encompassing the delusion of having turned into any animal. There have been reports of people thinking they have turned into a dog, a snake, a hyena and even a bee. It is incredibly rare. When Jan Dirk Blom, a psychiatrist at the Parnassia Psychiatric Institute in the Netherlands, searched through international records, he found just thirteen verified reports of people with the delusion of turning into a wolf in the previous 162 years.

I was intrigued but somewhat disturbed by this unusual disorder. Sharon and Rubén had shown me how easily one person's perception of the world can differ from another's, and Sylvia had opened my eyes to the hallucinations that we can all experience, but this felt so much more extreme. How can our brain be so dismissive of our human form? How can a person be convinced that they possess not arms and legs but claws or wings? What is it like, I wondered, to look into the mirror and see an animal staring back? And could it tell us anything about the way we think about our own bodies?

As Blom discovered, cases are few and far between, so I didn't expect to be able to meet anyone who had suffered from clinical lycanthropy. Nevertheless, I regularly checked in with specialist physicians and psychiatrists to see whether they knew of anyone who'd had the disorder. It quickly became apparent that clinical lycanthropy is not a condition in its own right, but appears alongside other more common mental illnesses, such as schizophrenia. Most doctors I spoke to said they had never come across it. One man who had was Hamdy Moselhy, chair of the College of Medicine and Health Sciences at the United Arab Emirates University. In fact, he is one of the few researchers in the world who has treated the condition more than once.

Hamdy's first encounter with clinical lycanthropy was back in the early 1990s, while working as a registrar at All Saints' Hospital in Birmingham, England. There he met a thirty-six-year-old man who had been behaving strangely for several years, ever since being arrested for wandering into the path of an oncoming car. His patient had been crawling on the floor, barking and eating vomit from the streets. He told doctors that he believed he was a dog and heard voices telling him to do things that a dog would do, like drink water from the toilet.[3]

"I'd never heard of this phenomenon in psychiatry," said Hamdy, when I first spoke to him about it. "I thought he might be pretending to feel this way in order to get away from a crime." He talked to his supervisor who told him to go and read up on lycanthropy. Keen to learn from past cases, Hamdy scoured the medical literature.

He discovered a description of a thirty-four-year-old woman who came to the emergency room agitated and tense. Suddenly, she started jumping around like a frog, croaking and darting out her tongue as if to catch a fly. Another case study described a woman who had the strange feeling that she was becoming a bee—she felt she was getting smaller and smaller.[4]

Late in 2015, Hamdy emailed me to say that he had a patient called Matar who had suffered from lycanthropy on and off for years—for hours on end he would be convinced he had turned into a tiger. Now, though, he had his condition under control and was happy to talk to me about it. "Would you like to come to Abu Dhabi and meet him?"

It is nine o'clock in the morning and already the thermometer in the car is creeping up to 111 degrees Fahrenheit. From the comfort of my air-conditioned taxi, I watch the gleaming sky-scrapers flash by my window. The gigantic brown and gold turrets of the Sheikh Zayed Grand Mosque—the largest in the United Arab Emirates—stand on the horizon. We travel west until we reach the outskirts of the city, where the grand buildings disintegrate into tiny rows of rundown shops. As we turn onto a five-lane highway lined with palm trees, quite suddenly the buildings disappear, as if we've reached some invisible border. The view on either side becomes a barren mix of desert dunes, the odd tree and an occasional sign for a distant camel racetrack.

The scenery stays this way for an hour.

"The people of Al Ain are pure, local village people," my driver, Amjud, says suddenly, waking me from my dune-induced trance. I look around and notice that the side of the road has become a little greener.

Its community may think of themselves as village people, but Al Ain is in fact the UAE's fourth biggest city, situated close to the border of Oman. It is sometimes known as the garden city, a reflection of its numerous parks and tree-lined avenues.

Down one of these avenues is Al Ain Hospital, where Amjud parks and I jump out. The hot air hits me like the opening of an oven, and so I walk quickly toward the nearest air-conditioned building. There I am met by Hamdy and Rafia Rahim, a softly spoken and fiercely intelligent specialist physician. As the three of us head back into the main hospital, I ask Rafia if Matar is well.

"He is fine," she says, "but he has been a little anxious this morning."

MATAR IS SITTING in a chair at the side of a wide, busy corridor. He is wearing a traditional kandura (a long white shirtlike garment) and a white headdress. He is in his mid-forties, but the dark circles under his eyes age him. He has a thick black beard speckled with gray and chubby cheeks heavily lined with wrinkles.

He gets up from his seat and looks at Hamdy, who greets him warmly.

"This is Helen," says Hamdy. I stretch out my hand and Matar shakes it gently.

We walk as a group through the hospital to a wing of empty offices. At the end of the corridor is a tiny study containing a single desk and four chairs. Hamdy asks us to take a seat while he goes in search of some water. Matar chooses the chair nearest the door, while I sit at a right angle to him. Rafia leaves us briefly to check on something in her office.

Alone together, I smile at Matar and thank him for coming into the hospital to see me. He stares at me, then tilts his head to the side, looking momentarily puzzled. I ask him how he is. Once again, he gives little indication that he understands what I am saying. I know that Matar isn't fluent in English, but I was under the impression that he knew a little. I smile and nod at the door. "I'll wait for Hamdy."

As we sit there in silence, I think back to what I already know about Matar. He was sixteen when he was diagnosed with schizophrenia. At the time, he had frequent admissions to the local psychiatry inpatient unit. Once, he had called the police because he believed that the UAE was under attack, after experiencing auditory and visual hallucinations of bombs exploding. Based on his call, the army was mobilized. He was later arrested for making false charges.

As an adult, Matar told his doctors that alongside his regular hallucinations, he had begun to turn into a tiger at night. He said he felt claws starting to appear from his hands and feet and that he would roar around the room. When it happened, he would lock himself in his room because he feared that if he went outside he might eat someone. He had told Hamdy how one day he felt himself turn into a tiger while having his hair cut; he jumped up from the chair and tried to bite the barber.

SCHIZOPHRENIA IS OFTEN SAID to be one of the most complex of all human disorders. It affects about one in one hundred people and common symptoms include paranoia, hallucinations, disorganized thinking and lack of motivation. Despite a strong genetic component (people who have immediate family members with schizophrenia have a much higher risk of having the disorder themselves), as well as clear environmental triggers such as trauma and drug use, we still do not know exactly why it occurs.

Some of the genetic research points to a mutation on chromosome 22, within a region known to be involved in the development and maturation of neurons. When researchers at the RIKEN Brain Science Institute in Japan grew neurons from stem cells obtained from people with this mutation, they found that fewer neurons grew, and those that did grow migrated over shorter distances, than did stem cells taken from people without the mutation.[5] It suggests that the mutation could cause abnormal growth and development in the earliest stages of life, which may affect how different neural networks within the brain communicate.

With such a wide constellation of symptoms, it has been difficult to pinpoint which neural networks are most affected. However, in recent years, it has been suggested that some symptoms of schizophrenia may arise from disruption to networks that allow us to distinguish between things that are generated by our self and things that belong to our external world.

We tend to give little thought to this distinction. Most of us instinctively know, for instance, that when we stick out our leg or tell a joke it is our own leg that's moving and our own words that we can hear. But the reason we are able to come to this conclusion is because our brain can predict the sensory consequences of our own actions, which gives us the feeling of being in control of the things we say and do. Since the late 1980s, Chris Frith at University College London and his colleagues have been developing a model for how this sense of agency arises and how it might explain some of the symptoms of schizophrenia.[6]

Let's use your leg as an example: give it a wiggle. To make this movement, your motor cortex—a region toward the top of the brain—sends messages to the muscles in your leg instructing them to move back and forth. According to Frith's model, at the same time, a copy of this message gets sent to other areas of the brain, which create a mental representation of the intended movement—a prediction of the consequences of this action. Once your leg wiggles, all the sensations that it creates—from the sight of your leg moving to the sensations that arise from the movement of your skin, tendons and joints—are compared to this prediction. If they match, you gain a sense of agency over the action.

These self-generated movements are registered in the brain less acutely than those sensations initiated by someone else. It's a clever adaptation—it means we don't jump out of our skin every time we touch our own arm as we might if someone else unexpectedly grabs us. In much the same way, when we speak, the brain appears to send a copy of the instruction to move our vocal cords to the auditory cortex. A few hundred milliseconds after we speak, our auditory cortex is dampened down. This doesn't happen when you hear someone else speak. It suggests that the brain predicts what sound you intend to make based on your vocal movements and compares this prediction to incoming sounds. If the two match, the sound is understood as your own, and somewhat ignored.

But if any part of this system goes awry, perhaps because of poor communication or bad internal timing mechanisms, we can no longer link our intentions with our actions and their predicted consequences. The brain is then forced to produce some other explanation for why these things are happening.

In 2016, Anne-Laure Lemaitre and her colleagues at the University of Lille, France, tested the theory that this is what is happening in schizophrenia with a simple experiment that you can try at home. All you have to do is remove your top, stretch your left arm up to the sky and, with your right hand, reach into your armpit and give yourself a little tickle. It probably has zero effect—it's really hard to tickle yourself. That's because our brain is predicting the consequences of the movements of our right hand and suppressing our reaction to them. The element of expectation and surprise—necessary for a good tickle—is gone. But when Lemaitre tested the ability of people with schizophrenic-like traits to tickle themselves with a feather, she found they were much more likely to report a ticklish sensation than a non-schizophrenic control group.[7] The results support the theory that people with schizophrenia are less able to predict the sensory consequences of their actions, which may lead to problems differentiating between sensations that arise from themselves and those that are externally produced.

We also see disruptions in the mechanisms that help us predict the sound of our own voice in people with schizophrenia, suggesting their brains cannot readily distinguish between internally and externally generated voices. It doesn't take a great leap of imagination to see how these disruptions could lead to a person concluding that they are not in control of their actions or that an internal monologue is coming not from themselves but from somewhere else.

Hamdy interrupts my thoughts, returning with small pots of water for all of us. He sits down next to me, shortly followed by Rafia, who takes a seat behind the desk.

Hamdy acts as translator while I thank Matar for coming to the hospital that day. He hadn't needed an appointment. He lives in a nearby village with his mum and sister, and had traveled in alone especially to speak with me.

I ask Matar whether he is happy to tell me a bit about his background, where he grew up, whether he has a partner. He thinks about the question for a second or two, and begins to speak softly, telling me that he has a wife. But almost immediately he hesitates. I have read that people who have suffered from lycanthropy can often show signs of shyness, so I turn to Hamdy. "Please let him know that he doesn't have to answer any questions he doesn't feel comfortable with."

Suddenly Matar grimaces, throws his head back and produces a strange sound. I'm momentarily startled, before realizing that he is sobbing. He looks up at the ceiling while his shoulders rise up and down. Rafia grabs a box of tissues and slips them across the desk. Matar dries his eyes and apologizes. He says the reason he is upset is because he has two children whom he no longer sees. One is fourteen years old and the other is eight, he thinks. He says he doesn't know exactly, because he hasn't seen them properly for a long time.

"My wife doesn't want me to see them at the moment," he says. "They live quite far away."

Hamdy turns to me and explains that Matar's wife took their children away after Matar started experiencing symptoms of lycanthropy, believing he might be a danger to them. I nod, trying to present some kind of understanding through my actions, if not my speech.

After a few moments, Hamdy asks Matar if he would like to

continue with his interview. He says yes, so I ask him how his symptoms began, and what they felt like.

"My schizophrenia started with visual hallucinations," he says. "I saw people coming and going that weren't really there. I could feel men and women and children grabbing at my legs and then falling to the floor."

His hallucinations grew worse over time. "It felt like people were starting to control my speech, that they could read my mind. They weren't allowing me to talk."

Suddenly Matar stops and looks at me strangely. He says something to Hamdy and jabs his finger in my direction.

I look at Hamdy.

"He says he's suspicious of you because you're British," he says.

"Why?"

Hamdy turns to Matar and asks him to explain his reasoning.

"We're all talking too much English," says Hamdy. "It's making him anxious."

The two of them chat for a while in Arabic. At the end of the conversation, Matar seems calm. He says he actually really likes the UK. He tells me that he got a scholarship to study at a British university, but that he needs to learn the language better. He says he'd like to study there one day.

He seems more at ease so I ask him whether he can explain what used to happen when he felt like he was turning into a tiger. Matar thinks for a moment and then points to his head and neck. "I feel it in my thoughts and in my body," he says.

He rolls up his sleeve and shows me his arm. He pulls at his thick black hairs, making them stand on end.

"When it starts to happen, all my hairs stand upright. The hair all over my body becomes erect. Then I get a spiky, itchy feeling over my body and my beard. It starts with a pain in my left leg, then it moves to my right leg, then to my arms. I start to feel an electric-like sensation going through my body. Then it feels like I want to bite someone. I can't control it, I just know that I am turning into a tiger."

He pauses and touches his throat, then looks directly at me and says something I don't understand in Arabic.

I glance at Hamdy, who looks puzzled.

"Matar says he has that feeling now."

ALL TOO OFTEN, the media are guilty of portraying people with schizophrenia as being violent. In fact, there is little scientific evidence for this at all. When Beth McGinty at Johns Hopkins Bloomberg School of Public Health and her colleagues analyzed news coverage from 1995 to 2014, they found that 40 percent of all news stories about mental illness focused on a link between mental illness and violence. That's highly disproportionate to the actual rates of violence among people with mental illness.

In the UK, for instance, homicides due to mental disorders peaked in 1973, and then declined to a rate of 0.07 per 100,000 people in 2004—the last year for which data was analyzed. That compares to total homicides, which increased over the same period and peaked in 2004 at 1.5 per 100,000 people.[8]

It's a dangerous misperception among reporters, the public and policymakers that mental illness is at the root of violence. Needless to say, sometimes it is: the high-profile assassination attempt on the American politician Gabrielle Giffords, for example, was carried out by Jared Lee Loughner, who was subsequently diagnosed with paranoid schizophrenia. But most acts of violence are the result not of the hallucinations and paranoia that accompany schizophrenia, but of anger and emotional issues, drug and alcohol use. "Most people with mental illness are not violent toward others and most violence is not caused by mental illness," says McGinty.

These thoughts comfort me. I look at Hamdy and Rafia for

direction. They both speak to Matar quietly. They tell him to relax, that there is no need to feel anxious in here, that we are all friends.

The room is silent for what feels like several minutes. Matar seems to be fighting some kind of internal civil war. Suddenly he grips his legs.

"Do you feel like you want to attack?" Hamdy asks, breaking the silence.

Matar looks up at him. "How did you know that? Are you reading my mind?"

Hamdy assures him that he cannot read his mind, and that he is just asking how he is feeling.

Matar looks at him with suspicion. Then he says something in Arabic that makes Hamdy laugh softly.

"What's going on?" I ask.

"Matar asked me if I am really the Hamdy he knows. He thinks I might be an impostor. He says the Hamdy he remembers is really fat."

Matar nods. "The Hamdy I know is obese," he says.

I raise an eyebrow at Hamdy. "No, he's right," he says, smiling. "I haven't seen Matar in person for a year or so and when I last saw him, I really was obese."

Hamdy explains to Matar that he's lost a lot of weight recently and that surely he recognizes both him and Rafia.

"The Hamdy I knew was more kind," says Matar.

Hamdy smiles and chats to Matar for a little while longer. He asks if he wants to stop or carry on. Suddenly Matar's shoulders relax, and his eyes become more focused.

"Yes, let's continue," he says.

I take a deep breath and ask Matar what it is about his delusions that made him feel like a tiger, rather than, say, a cat or some other animal.

"I feel like you are eating my legs, like a Kentucky drumstick,"

Matar says, ignoring my question. "You feel like a lion to me, I want to attack you before you attack me."

My gut clenches. There is no getting around it, Matar is clearly in the midst of a terrible relapse. He takes a sudden intake of breath and looks down into his lap, and a deep and incredibly realistic growl rumbles from his mouth.

My pen hovers above my notepad and I find myself imagining what a predator and prey might do in this situation. Hamdy is sitting on my left and the door is to my right. But I don't want to move; I don't want to startle him. Matar has both hands clenched on the tops of his legs and his fingers have begun flexing as if they have claws. The growling is directed at me. When Hamdy tries to speak, the growls turn toward him.

"Are you wanting to attack us?" asks Hamdy.

"All three of you," says Matar.

Both doctors glance at each other. They start to talk at once in English and Arabic.

"Relax, Matar, it's okay. You know who we are and why we're here. You wanted to come and talk to Helen about your condition, remember?"

Matar nods. He seems to be trying to fight the urge to attack. He takes a few deep breaths and suddenly becomes quite lucid again. He says he needs a cigarette. Rafia slips from behind the desk and helps him out of the room.

With Matar gone, I turn to Hamdy and ask his opinion about what has just happened.

"I don't think he has taken his medication," Hamdy replies. Matar usually takes a mixture of antipsychotics, antidepressants and antianxiety drugs, he says, which help control his symptoms. "Something must have happened to make him stop taking them. I don't think we are safe in this room."

I agree, and suggest we end the interview here. Hamdy disagrees; he says we should just move to a bigger room.

"You should sit by the door so you can run out if you need to."

I am feeling terrible about the possibility of making Matar's relapse worse, but follow doctor's orders. I get the feeling that this is a rare opportunity for Hamdy and Rafia to find out more about the condition, to understand it better. We walk to a large seminar room, set out with rows of chairs.

While we wait, I ask Hamdy why it is that Matar's schizophrenia has manifested itself in this rare belief that he can turn into a tiger. Why does this happen to him, but not to others with the condition?

Hamdy says that is the million-dollar question. "There's something different going on," he says. "People with lycanthropy see their bodies not as human, but as animal. We have to ask, 'How is that possible?'"

We may not be able to find out the answer from studying people with lycanthropy—there are just too few around—but that's not to say we can't make some inroads. You don't need to be suffering from lycanthropy to feel as though your body is changing shape, or altered in some way. There are many strange disorders in which people feel like their limbs are unwanted, are present when they're not, or have grown smaller or larger. Some of these can give us clues to what might be happening to Matar. But to find out more, we need to travel back to 1934, where a young man is lying in an operating room, head shaved, brain exposed—and wide awake.

Wilder Penfield gripped the tiny electrode and lowered it onto the surface of the young man's brain. He pressed a button and a tiny current ran through the metal rod, giving the surface of the brain below a small jolt.

"What do you feel?" he asked his patient.

"I feel a tingling sensation on my jaw," he said.

Penfield's assistant made a note of the result and placed a marker onto the area of the brain that had just been stimulated. Penfield moved the electrode a fraction of an inch and began the process again. This time the patient felt a sensation of being touched on his upper arm.

We met Penfield in Bob's chapter, when he was stimulating an area close to the hippocampus to produce flashes of memories in his patients. This time, he was trying to identify which areas of his patient's brain were causing epileptic activity and needed to be removed, and which healthy tissue he should avoid. He would often start such an operation by identifying the central sulcus, a prominent indentation at the top of the brain that separates the frontal lobe from the parietal lobe. Just in front of this landmark is the primary motor cortex, a strip of tissue that contains cells that travel down into the spinal cord, where they connect with motor neurons that terminate in our muscles. Just behind the central sulcus is the parietal lobe, which contains a similar strip of tissue called the primary somatosensory cortex. This contains cells that receive information about tactile sensations from all around the body. When Penfield stimulated the primary motor cortex, his patient would feel the sensation of movement of a specific muscle. When the somatosensory cortex was stimulated, he would feel the sensation of being touched.[9]

Using hundreds of these operations, Penfield was able to create a cortical map of the "body in the brain." In doing so, he discovered that the body is mapped within these strips of tissue in a familiar order, meaning body parts that are adjacent in real life were adjacent in the brain. So an area of the somatosensory cortex that provoked the sensation of touch on the upper leg would be close to an area of the brain that provoked the sensation of touch on the lower leg. This in turn would be next to an area responsible for the ankle, the foot, the toes and so on.

Penfield illustrated these body maps using what is now known as a homunculus—an image of a grotesquely squat man with extraordinarily large hands, fingers, lips and tongue. The homunculus is disfigured because it represents the size of the brain area devoted to a given body part—and each area is proportional not to the body part's physical size, but to how richly innervated it is by muscles or sensory nerve endings. For instance, the sensory homunculus has disproportionately huge lips and hands because they are extremely sensitive to touch and therefore take up a lot of room in the brain. Areas such as the torso and upper arms are tiny on the homunculus because they have fewer nerve endings and therefore take up less room.

These maps are important in giving us a sense of what our body looks like and where each part of it is at any given moment. This may sound strange—perhaps you think you know what your body looks like because you can see it—but visual stimuli are not the only way you sense your body.

Close your eyes and stretch out your hand. Now touch your nose. You can do it despite not being able to see any of your body. And that's because embedded within the brain is a model of what your body should look like—an internally generated image that scientists sometimes call our sense of bodily self. In order to create this image, Penfield's motor and sensory maps work together with what's called a proprioceptive map, which processes information about your joints and movements. These maps aren't static—every second of the day they are being updated to give you a fluid sense of where your body is, what it feels like and what it is doing. If you start to put on weight, for instance, your visual sensations of a bulging midline and internal sensations from the skin and muscles both contribute to update your brain's internal body schema. It's not clear exactly where your final body image is produced, although there is some evidence to suggest that the superior parietal lobe is involved (stroke patients who damage this area sometimes fail to recognize their limbs as their own). What we do know is that when all these body maps are communicating with one another, it produces the feeling of owning a body that matches the physical reality. The problem is that sometimes this system goes awry, and that's when we start to feel a little out of sorts.

Take, for example, the phantom limb—a phrase introduced in 1871 by the American neurologist Silas Weir Mitchell. People with a phantom limb experience an amputated limb as still present and sometimes painful. After Lord Nelson lost his right arm during the Battle of Santa Cruz de Tenerife he referred to the subsequent sensation of pain in his missing limb as "proof of the existence of the soul"—for if an arm can survive physical destruction, he said, then why not an entire person?

We now know that this is not proof of a soul, but of something else quite extraordinary—neural plasticity, or the brain's ability to reshape itself throughout our life. When someone has their limb amputated, areas of the brain that once received input from the amputated limb are now neglected. The brain doesn't like to waste precious real estate, so when one limb is erased, the rest of our body image quickly expands to take its place. This is why phantom limbs occur—brain areas that once processed touch to the arm, might now be taken over by neurons that process information about touch to the face. This can lead to the sensation that an amputated arm is being touched when in fact it is the face that is being touched.

Often these phantom limbs become painful; a phantom arm might feel like it is frozen or clenched in a fist. It's likely that motor regions of the brain are still attempting to send commands to the missing limb but are getting nothing back. These mixed messages make it seem as if the phantom limb is paralyzed. A simple trick can help relieve this pain almost instantaneously. The person sits with a mirror between their remaining limb and their phantom limb. Looking in the mirror provides an exact replica of the limb where the phantom is felt. Unclenching the fist, or moving the real limb, gives the impression that the phantom limb is making the same action. In this way, people can soothe pain or even make their phantom limbs disappear altogether.

You don't need to have lost a limb to discover what it's like to incorporate a phantom limb into your body image. Take a blown-up rubber glove and two small brushes and place the rubber hand on the table in front of you. Conceal your own hand with a piece of wood or cardboard. Now get a friend to brush the rubber hand repeatedly at the same time as brushing your real hand. Once the illusion kicks in you should start to feel like the rubber hand belongs to you, and that you can feel the brushstrokes directly on it.

This is the most famous example of how easily our body image can be altered, but there are many others. In 2011, Vilayanur Ramachandran at the University of California and his colleagues wrote about a new condition that they called xenomelia, in which otherwise rational individuals are overcome with the desire to amputate a healthy limb. Ramachandran's first subject was a twenty-nine-year-old man who recalled a strong desire to amputate his right leg from around the age of twelve. He said the leg made him feel "over-complete" and he simply wanted it gone. He readily acknowledged that these feelings were not normal. A month after first visiting Ramachandran, he poured dry ice over his lower leg, forcing surgeons to perform an amputation.

Many doctors claimed this was just a cry for attention or a result of psychological trauma from early exposure to an amputee. But Ramachandran argued that it was much more likely to have an identifiable biological mechanism in the brain.

"When we ask these patients to draw a line across their limb for where they wish it to be amputated, then ask them to draw it again a month later, the line is in the exact same place," he told me at the time. "It was just too specific to be down to an obsession of some kind."

To prove his point, he worked with the neuroscientist Paul McGeoch, also at the University of California, to analyze four people with xenomelia who wished to have one of their legs amputated. The experiment was simple: they scanned each patient's brain while touching their legs.

The results were startling: When subjects were touched on their "normal" leg or above the line of amputation on their unwanted leg, there was a significant burst of activity in their right superior parietal lobe. When they were touched on their unwanted limb, activity in this area didn't change. The team say that the right superior parietal lobe is an area of the brain that is ideally positioned to unify disparate sensory inputs to create a coherent sense of having a body. They propose that xenomelia occurs when someone finds themselves in the unnatural situation in which they can feel their limb being touched, but that sense of touch is not incorporated into their body image. The result is a desire to amputate the seemingly alien limb.[10]

Interestingly, something similar might explain why transgender people often feel at odds with their anatomy. Recently, Laura Case at the University of California and her colleagues recruited eight people who were anatomically female but identified as male with the strong desire to have male anatomy. They also recruited a group of non–trans women as a control. To find out if there were any significant differences in the way their brains processed information about sexualized body parts, Case and her team scanned each participant's brain while they were tapped on the hand or breast. As you would expect, stimulation of the hand and breast caused the corresponding areas of the parietal lobe to light up with activity. But in the trans group this activity was significantly lower when their breast was touched compared with when their hand was.[11]

We have a chicken-and-egg problem with both these studies— it's impossible to tell whether the brain differences are a cause or a consequence of a lifetime's aversion to a specific body part. Nevertheless, both experiments clearly demonstrate the importance of our internally generated body image, and specifically the parietal lobe, in helping us achieve that image. But can it also help to explain the presence of lycanthropy?

There's one early sign that it might. In 1999, Hamdy came across a fifty-three-year-old patient who suffered from epilepsy and severe depression. For some time, she found it difficult to escape the belief that she had claws growing out of her feet. A brain scan showed a loss of tissue in one side of her parietal lobe. It was the first hint that, when people with lycanthropy report their bodies changing shape, they may be genuinely perceiving those feelings.

We also know that people with schizophrenia show a higher susceptibility to body illusions such as the rubber-hand trick mentioned earlier. Brain scans suggest that this might be explained by a stronger reliance on sensory information from vision and movement over weaker stored body representations. It could suggest that in some extreme cases a visual hallucination of claws or an animal's face may be more easily incorporated into one's sense of body image.

Unfortunately, Matar's brain scans have not yet pinpointed any abnormalities. That's not to say there aren't any. Several doctors I have spoken to about the condition think that the unexplained nature of lycanthropy, as well as schizophrenia in general, might be revealed with better neuroimaging techniques with higher resolution than currently available.

"There are many hypotheses and partial solutions, but we need to scan more patients and do bigger studies before we can make any conclusions," says Hamdy. "So for now we continue to treat Matar's schizophrenia and hope that this also helps his belief that he can turn into a tiger."

Back in Al Ain, Rafia comes into the room followed by Matar and three young doctors. During his cigarette break, she had learned that Matar's mother was away with his sister in India. His sister had started to show signs of schizophrenia, and was being tested at a specialist hospital. Rafia seemed to think that Matar had stopped taking his medication, and his anxiety about his mother being away had exacerbated his symptoms.

Matar walks across the room, seemingly in better spirits. He sits in one of the front seats. "I'd like to carry on," he says, looking at me.

I smile, thank him and ask again if he could tell me how he knows he is a tiger when he gets these feelings, rather than any other animal.

This time he answers immediately and fluently.

"I don't know why I'm a tiger, I just know I am. I hear lots of voices around me, telling me I'm no good. They laugh at me. They tell me I'm rubbish, that I'm not good enough to be human. Some days it feels like there is a lion around me. Sometimes it is attacking me, grabbing the back of my neck. I can't move from the pain. I can see the blood coming from my body where I've been attacked."

"Do you ever feel like you can defend yourself?"

"No." He shakes his head. "I can't defend myself from the lion as it's much stronger than me so I feel like I have to attack first."

"How long does it last?"

"Sometimes only a few minutes, sometimes hours."

Hamdy interrupts. "Have you been feeling like this recently, or has it just happened today?" he asks.

"It started last night," says Matar. He looks upset. "I was in bed and I felt it coming on, so I locked the door, put a towel over my head and wrapped myself up in my sheets so that I couldn't move my arms or get free."

He says that once, when he could no longer resist acting out his urges, he attached cement blocks to his shoes to make his feet too heavy to move.

"I just want to stop myself from hurting anyone."

"Have you ever looked in the mirror when you've felt like a tiger?" I ask.

"Yes," he says. "I've looked at my reflection when I've felt like a tiger and I see two things. I see myself as a tiger and I see a lion that's catching hold of my head and my neck. I can't rationalize it. It's very frightening."

Despite his behavior today, Matar is not thought to be a danger to anyone. His medications generally allow him to function well in society and live safely in his local community.

"We're happy to have him live at home. He has his family and a community nurse to look out for him," says Hamdy. "It's different here than in the UK—there's a big emphasis on family helping care for people who are ill."

I turn back to Matar. "Is there anything you can do to help prevent your delusions from occurring, other than take your medication?"

"I always wear white," he says, pointing at his white long-sleeved robe and headdress. "It's calming. White feels like a peaceful color to wear, it helps when I have these weird feelings."

Suddenly the atmosphere changes again as Matar laughs out loud. He stretches his fingers out wide and bends them at the knuckle. He lowers his head and slips off his shoe. Then he grabs his left leg and grimaces in pain.

Suddenly, the snarling starts again.

"I think we should leave," says one of the doctors sitting beside me. Another asks Matar if he would like some medicine to help with his anxiety. He nods and with barely a glance walks out of the room and is gone.

This is the point where I'd love to add that Matar is now well, that a mixture of drugs and psychotherapy has helped him banish his delusions. Unfortunately, that's not the case. A few months after I returned home, I emailed Rafia and asked her to translate a note from me to Matar, thanking him for coming to meet me. I wanted to know how he was feeling. Rafia replied quickly. She said that Matar's behavior on the day of my interview suggested a fairly severe relapse and that he continued to have repeated admissions to the hospital. So far, she said, he has not returned to an acceptable level of functioning.

Matar's brain may be unique, but there are many lessons we can learn from such extreme cases, not least the incredible power of family in ensuring the health of those who may otherwise find themselves in permanent care. If you place people with lycanthropy side by side with others who experience perceived body alterations, though, what emerges is a picture of the brain working tirelessly to create a sense of something that most of us take for granted every day: a body that feels like our own.

LOUISE

Becoming Unreal

think it's time to ask you a question.

Who are you?

It seems simple enough. There are many ways that you could answer. When we think about who we are, it's often from the perspective of others. I might think about myself as a journalist, a daughter, a friend, a wife, a James Bond fan. But what else constitutes "me"? As Matar showed us, part of "me" is my body. Mine is fairly tall and slim. It has big feet.

But another aspect of me is also present inside that body. I am made up of emotions, memories, thoughts, opinions and bodily sensations that all carry a feeling of "being mine." The body I see in the mirror changes day by day, but the person inside that body is constant. Scientists like to call this our sense of self. This attachment to our thoughts and sensations seems like something that should be fairly permanent. Yet that is not always the case.

In his diary, the Swiss philosopher Henri Frédéric Amiel

describes a peculiar feeling: "I find myself regarding existence as though from beyond the tomb, from another world; all is strange to me; I am, as it were, outside my own body and individuality; I am depersonalised, detached, cut adrift. Is this madness?"[1]

These feelings were later defined as depersonalization disorder—a condition in which a person feels detached from themselves, as if their external and inner worlds have become unreal, as if, as Amiel put it, they are no longer tied to their own mental experiences. Some people describe it as feeling like they are watching a movie of themselves, or that their perception of the world has decreased in quality.

You might have had some experience of this yourself—it's thought that brief episodes of mild depersonalization are common among the general public, typically at moments of high stress or fatigue. That spaced-out feeling you get when jet-lagged or hungover could be seen as a transient experience of depersonalization, for instance. A number of drugs, including ecstasy, have also been associated with the onset of such feelings.

Depersonalization can appear suddenly without any apparent trigger, or after severe stress or childhood trauma. There are some theories that it is a protective mechanism, that in the face of extreme danger our sense of self can detach itself from what's happening in the world, in order to remove itself from the stressful nature of what's going on. This idea struck a chord with me because it brought back memories of a fleeting feeling of depersonalization that I had experienced, one morning driving to the dentist.

The road was wet and the gravel and fallen leaves had washed to the bottom of a T junction near my house. As I braked to meet the end of the road, my wheels skidded and I drove full speed into the oncoming traffic, spun and crashed into a lamppost. The entire episode not only felt as though it was happening in slow motion, but also felt like it was happening to someone else. As soon as the brakes stopped working, I remember a distinct feeling of detachment from my body, like it didn't quite belong to me. I remember thinking back to my driving lessons and whether my driving instructor had told me anything about what to do in this situation. I decided he hadn't, and then proceeded to feel angry about his lack of foresight. Then I recollected someone pumping the pedals during a skid in a film I had once watched, so I tried that. After that didn't gain me any traction, I remember searching the oncoming traffic to work out which car I was likely to hit. I felt like I was watching myself get closer to the end of the road. I remember thinking about whether there was any way I might warn them and tried to look apologetic as I neared the unsuspecting driver. After hitting the first car and spinning 180 degrees I remember thinking it was strange that the air bag hadn't worked, and then trying to figure out how best to position myself to minimize the damage for the second impact. It was almost as if my physical body was not quite "me." Like I was watching it all happen from somewhere inside my own head. This slow motion catastrophe ended as my car eventually came to a halt across the road. If these were feelings of depersonalization, they were short-lived: as soon as the lamppost buckled the front of the car, the pain down my shoulders dragged me back into myself. But for some this feeling of depersonalization, this detachment from your sense of self, is a permanent way of life.

I pull into a tiny back street, lined with colorful terrace houses. The rain is lashing down and the cobbled road is so small even my Mini is struggling to make a three-point turn. I'm in the seaside town of Brighton, down on the south coast of the UK.

I lock up the car and dive for cover as the door to the nearest house opens and a small child grins out at me. "Are you Helen?" he says.

I follow the child into his house. "Hello?" I call. Louise suddenly appears at the top of the stairs and beckons me up. From the living room, two more kids stare out at me.

"Hi," she says brightly. "Sorry, it's a bit hectic around here. They're not all mine. Tea?"

This was not the same Louise that I had met a few years previously. Back then, sitting in the Tate Gallery in central London, she had seemed distracted, wary, exhausted. She had gazed blankly at the people around us. She had said that she felt like she was in a play and that everyone around her, including me, were the actors. She felt completely detached from the world. "I can hear myself speaking to you," she'd said. "I know rationally that this is my voice, but it doesn't feel like it's mine. None of this feels real."

Now, standing in her kitchen with the kettle boiling, Louise looks like a different person. Her blond wavy hair has changed to a deep shade of brown for a start. But the real difference is in her eyes: a year before they had seemed distant and unsure of the world, now they are clear and focused and she is smiling, confident and cracking jokes about the havoc in the next room. Louise pours out two cups of tea and then points me back down the narrow flight of stairs.

When we are out of earshot, Louise says, "When I get depersonalized now, I don't panic, I just tell myself, it's not real, it's just my brain, nothing has changed, this is still my arm, this is still my house, and just get on with the day."

She opens what I had assumed was the garage, and welcomes me into a brightly decorated tiki bar. "I converted it a few years ago," she explains.

The room is filled with glasses and candles and strings of

lanterns, with tribal masks and leis hung on the wall. I settle myself on a barstool and ask Louise to take me right back to the beginning.

SHE WAS EIGHT YEARS OLD and off sick from school the first time it happened.

"I woke up that morning, and suddenly felt like I'd been dropped into my body," she says. "It's really hard to describe, but it was like I was just born. Everything around me felt new. It's like you are completely different to what you were a second ago. A completely different you. You're suddenly really aware of where you are and who you are, and everything around you feels foreign . . ."

She pauses. "Everything about yourself and everything around you feels alien. You know rationally that it can't have changed, but it's like you're walking around in this world that you recognize but no longer feel. It's like this unshakable sense of detachment from your body and the world.

"Agh!" she groans. "It's so hard to explain."

Louise, like many with depersonalization, has great difficulty describing her state of mind; no eloquent metaphor seems to sum up her feelings accurately. She tries one more time: "It's like you're watching the world but are no longer part of it."

Her first few bouts of depersonalization were brief. "As a child, it only happened for a few minutes at a time," she says. "I would panic and run to be around others, but I never talked about it with anyone."

"Why not?"

"I don't know, I just thought it was really weird. I didn't want anyone to think I was crazy."

This is where depersonalization differs from schizophrenia. This disturbing feeling that your experience of yourself and the world around you has changed is not accompanied by any kind of psychosis. The people who suffer from this disorder never lose the ability to distinguish what is real and what is not.

"You never truly believe that you're part of some alternate reality, but that's part of the problem," says Louise. "You know rationally that these weird feelings that you're experiencing can't be real, that the world you are in hasn't suddenly changed, but it still feels like it has. That's why it's so frightening. It's worse than being away with the fairies—it's like being a sane person who is mad."

THE FIRST TIME that Louise's depersonalization became really bothersome was at university. She had been suffering from a migraine when suddenly her world became distant, separate from her body. She was floating around, she says, in a world that she was no longer part of. This feeling stuck around for days at a time.

"Then it started lasting a week, and then longer. Eventually it just set in and wouldn't lift. Finally I had to leave university. I became permanently anxious, like that feeling you'd have if you tipped backward on a chair and are about to fall. I'd feel like that all the time. I couldn't stop thinking about how weird everything felt. I thought I was going mad—it was just terrifying," she says.

Despite this inner turmoil, it was not immediately obvious to those around Louise that there was anything wrong. She knew rationally how she should be behaving, so her actions appeared perfectly normal to others. Yet she spent years feeling isolated, frustrated and fearful. After countless visits to doctors who shrugged their shoulders in response to her weird concoction of symptoms, Louise became depressed, terrified of the constant anxiety attacks that would coincide with these bouts of unreality.

"Sometimes," she says, "when it's at its worst, I can't have any noise in the house. When you're in this state, everything around you feels like it is screaming at you to get noticed. But at the same time your whole world seems like it's happening to someone else, someone you're not in control of. It's like walking through tar. It's exhausting."

"And there's no way of ignoring it?" I ask. "By concentrating on the rational part of your brain that's telling you everything is okay?"

"No," she replies. "Saying 'think positively' is like trying to fix a leg that's been blown off with a plaster."

She's quiet for a while. "Have you seen that painting by Edvard Munch?" she says, suddenly. "With the face screaming against an orange sky? Some say it is about depersonalization."

In the 1800s, Munch created four pieces of artwork called *The Scream in Nature*. The pieces are made from oil, pastels and crayon and each shows a ghostly figure with a skull-like face looking out from the canvas, its mouth wide open and hands either side of its cheeks. The sky behind the figure is made of red swirls and there appears to be water in the distance. Two people stand close by, seemingly oblivious to the figure's turmoil. Expressionists of the time often placed emphasis on painting their inner feelings and emotions rather than a realistic image. "It's not the chair that should be painted," said Munch, "but what a person has felt at the sight of it."[2]

In a poem Munch painted onto the frame of *The Scream in Nature*, he said: "I was walking along the road with two friends—The Sun was setting—the Sky turned blood-red. And I felt a wave of Sadness—I paused, tired to Death—Above the blue-black Fjord and City Blood and Flaming tongues hovered—My friends walked on—I stayed behind—quaking with Angst—I felt the great Scream in Nature."[3]

"That painting makes perfect sense to me," says Louise. "The person and the landscape are screaming at you. It's exactly like depersonalization—when it happens you can't get any peace. It's not only the outside world that seems strange, but your internal one too. Everything you're familiar with becomes alien. You're detached from everything—even your memories. Memories of things you've done suddenly don't feel like they belong to you. It robs you of your past. It takes away the core of who you are."

"Your memories don't feel like your own?"

"Yeah, you just feel separate from everything that you thought you were. Your memories, your voice. I mean I know this is my voice and these are my memories, but when I'm in that state, they don't seem like mine. I know I'm controlling what I'm saying but it's like I'm in a film, like it doesn't belong to me. It's like I'm on my own in the center of everything and no one else is real. It makes you feel very separated and lonely from everything, like you're the only person in the world that is really here."

A few years before we first met, Louise ended up in the hospital. She had just given birth to her second child.

"I'd had illnesses running up to the birth and I'd felt really odd throughout the pregnancy," she says. "When you're in charge of someone else and you're not feeling like you're in charge of you, it's just the worst. After the birth, it was the first time I really relaxed. But then I went into the shower and suddenly the weirdness hit me again. I had this massive panic attack. My whole world closed in on me and everything went black."

Two months passed, which Louise says were a blur. "I can't really remember anything between giving birth and having to go back to the hospital. I just couldn't manage anymore. It had come to the forefront of my mind and I was so overwhelmed by it all I couldn't think of anything else."

Her husband could see there was something wrong, but it wasn't obvious to him what it was. "Everyone kept asking about depression and whether I had suicidal thoughts. I told them that any bad thoughts were only there because I wanted these weird feelings to stop. I had a little baby and I wanted to get on with my life. It was just one massive nightmare. It was like being in hell. I wouldn't wish it on anyone."

After first meeting Louise, I paid a visit to an online forum for people who suffer from depersonalization to learn more about the disorder.[4] I browsed through some of the posts. One member said he had an acute feeling of forgetting who he was and the way that humans live. "I feel like I'm just an alien from another dimension pretending to be human the best I can," he said. "All my memories are there but it's like I can't trust them, my brain won't accept and assimilate them." Another described themselves as "a frame, not even a shell. All that was me is no longer there." Some appeared to be confined to their houses, unwilling to spend their day interacting with people who didn't feel part of their world. One regular contributor did the opposite: he walked ten miles a day, yet still felt nothing. "I'm so fucking numb I can go anywhere, do anything and I won't feel a damn thing."

There seemed to be a common theme of emotional numbing of one sort or another. I ask Louise whether she too felt she lacked emotions toward people and her surroundings.

"When I think about it rationally, I do have an emotional attachment to others, my parents or my husband, for example, but if they are around and I'm feeling depersonalized then it's like the room is your play, and the space in front of you is your set. They're all just actors. So at that point I don't feel any particular attachment or emotion toward them or the things around me."

I'm struck by this peculiar paradox: Louise and those in the depersonalization forum all talk about an emotional numbness and disconnection with themselves and the outside world, yet all suffer strong feelings of subjective distress over this weirdness. As in Munch's painting, they say that the world seems to be screaming at them, yet they feel like they're not part of it. How is it possible to feel nothing and everything all at once?

The answer, it seems, begins with a rather unusual murder.

IN THE SUMMER OF 1921, William Hightower told a newspaper reporter that he had been digging in the sand at a popular spot at Salada Beach in California, hoping to find some bootleg whiskey that he had been told was buried there. While digging, he said, he had uncovered a black prayer scarf that he believed belonged to Father Patrick Heslin—a local priest who had gone missing a week earlier and who had been the subject of several ransom notes.

Eager to claim a reward for having found Heslin, Hightower returned to the spot with a reporter, who brought with him a team of police. They started digging. One of the policemen told Hightower to be careful where he was digging as he might strike the face of anyone buried below. Hightower told the policeman not to worry as he was digging at the priest's feet. Hightower was arrested, and Father Heslin's body was uncovered.

The *San Francisco Call and Post* arranged for Hightower to be hooked up to a new piece of machinery, the cardio-pneumo-psychograph, invented by John Augustus Larson, which was quickly renamed the "lie detector" by local press. Larson's technology incorporated a test for blood pressure with tests for skin conductivity, pulse and respiration. He believed that fluctuations in these bodily functions were an excellent means of detecting guilt. Hightower was the first person to trial the technology. On August 17, the newspaper's headline read, "Science Indicates Hightower's Guilt." Police later found in

Hightower's hotel room the pistol used to shoot the priest, the typewriter used to write the ransom notes and some sand.

While the wild gyrations of the lie detector were never reliable enough to be fully accepted by the scientific community, it was one of the first demonstrations that our unconscious bodily functions are linked with our thoughts.

Have you ever been told to "go with your gut"? Or to "follow your heart"? We talk about leaning on our bodily sensations all the time, but there's more to this than a figure of speech. Take your heartbeat, for instance. Think about it now. Can you feel it pumping softly against your breastbone? Perhaps it's pounding. Or maybe you are struggling to feel it at all. Take a moment and try counting it without touching your chest or feeling for your pulse. Is it more difficult than you thought? When asked to count their heartbeat for a short time, one in four people are off by about 50 percent.

Our ability to sense the physical condition of our body is called interoception. You should be familiar with this concept even if you don't realize it, since with the exception of a few, most people can sense whether they are cold or hot, where a pain is coming from, whether they are thirsty or hungry. These are all interoceptive feelings.

Scientists tend to use the heartbeat test as a way of measuring our interoceptive capability. Each of us has a different capacity for such bodily awareness and we now know that this ability is intimately connected with our thoughts, feelings and social behavior. People who are more in tune with their heartbeat, for instance, are better at reading their own emotional feelings. People who are better at interpreting their own feelings are also subsequently better at interpreting the emotions of others. People who have greater interoceptive abilities also make better decisions based on subtle cues in their environment, and can make intuitive choices more quickly. They can judge the passing of time more accurately and also perform better in tasks that require them to divide their attention.

One noteworthy example of how interoception affects our thoughts comes from the journal *Social Cognitive and Affective Neuroscience*, in which the Argentinian neuroscientist Agustin Ibáñez describes "the man with two hearts."[5] The gentleman in question had heart disease, and so Ibáñez replaced his failing heart with a mechanical pump. Unfortunately, his patient disliked the sensation of his second heart, which was implanted just above his belly button. The mechanical throbbing sensation made it feel as if his chest had fallen into his abdomen, he said. But, interestingly, the sensation of his second heart also affected his behavior. His new heart did not react to external events like his original heart. Before his operation he had no problems empathizing with others. Now that his mechanical heart ruled, he had problems reading other's motives, he lacked empathy when looking at painful images and even had difficulty making decisions.

It supports a theory first proposed by William James in the nineteenth century, that we may be able to register what's happening in our external world in an intellectual and rational manner, but it is an awareness of our *body's* reaction to the world (our beating heart and sweaty palms) that conjures up our rich emotional response to that world.

Some of the most influential work in this area has been carried out by the Portuguese-born neuroscientist Antonio Damasio, who describes emotions and feelings as two separate things.[6] Emotions, he says, are the brain's reaction to certain physical stimuli. For example, if a rabid dog starts barking at us, our hearts might race, our muscles contract and our mouths go dry. This emotional reaction happens automatically. Our brain then attaches value to this emotion—is it rewarding, strong, negative?

Our feelings occur only after we become aware of the physical changes to our body and start to form a conscious representation of that emotion—something we then attach a word to and call our feelings.[7]

You can test this idea right now. Start by clenching the corners of your mouth. Slowly pull each side up. Then a little bit more. Now open your mouth slightly. Squeeze your cheeks up toward your eyes and there you have it, you're smiling. Stay like this for a moment. Do you feel better? You should—scientists have shown that the physical act of smiling actually makes you feel happier. According to Damasio's theory, the brain notices the muscle movements associated with smiling, attaches all the positive values it knows are associated with this reaction and creates the feeling of happiness.

Recent studies have suggested that the brain region responsible for integrating all the information coming from our internal sensations is the insula—a fold that lies deep in the center of the brain. One promising theory suggests that information coming from the body is collated and integrated in the back and middle areas of the insula, before being re-represented by the front, or anterior, insula, which generates the feeling that pops up into our consciousness.

"The anterior insula is the area of the brain that forms a default setting of "'this is me here and now,'" says Nick Medford, a consciousness expert at Brighton and Sussex Medical School. Medford spends much of his time placing people in brain scanners and showing them pictures of grotesque surgery, filthy bathrooms and cockroaches—images designed to elicit aversive reactions. When we look at these kinds of highly evocative stimuli, the insula lights up with activity. When Medford showed fourteen people with depersonalization these images, however, he found a startling lack of activity, specifically in the left anterior insula, compared with people without the disorder. The study also showed some evidence that a region called the ventrolateral prefrontal cortex might also be involved in inhibiting the insula's response to the gruesome images. This area is known to help keep our emotions in check; in people with depersonalization disorder, it seems to be overactive, or too controlling.

Ten of the fourteen people with depersonalization in Medford's study had their brains scanned again, four to eight months after receiving a drug used in mood disorders. Those who showed increased activation of the insula were the same people whose symptoms had improved. Those who showed improvements also had decreased activation in their ventrolateral prefrontal cortex after taking the drugs, whereas it was still active in those whose symptoms had persisted.[8]

Medford figured that if people with depersonalization had a suppressed neural response to the world, then this would also be apparent in their body's autonomic reaction to stimuli. In this context, the word "autonomic" refers to the things that happen in the body that we cannot control (the basis for Larson's lie-detector test). Medford focused on skin conductance—a way of measuring when the skin momentarily becomes a better conductor of electricity as we become aroused. When this occurs, sweat glands become more active and conductance increases. Skin conductance is one of neuroscientists' favorite tools since it allows us to gain an objective insight into our emotional responses: you can't fake a sweaty palm.

And neither could those with depersonalization. No matter how grotesque or displeasing a picture, people with depersonalization showed few signs of their body reacting.[9] Somehow their body's automatic response to the outside world has been dialed down, and is not integrated into subjective feelings about themselves or the world around them. But why does this create the sense that they do not own their own voice, or that the world has become unreal?

It turns out the explanation might be related to something we have already come across—the idea that the brain makes sense of the world using predictions. As we saw in Sylvia's chapter, the brain does not process every single aspect of the sensory input it receives from the body and the outside world; instead it makes a "best guess" at what these inputs might mean. If it makes an errant prediction, then it either updates future predictions or creates a perception of the world that better suits the input it is receiving.

This prediction model might also explain depersonalization disorder. When everything is working well, the brain makes predictions about things that are happening inside the body and these match with the actual signals it receives. This culminates in feelings that can be attributed to "you." If there is something wrong, however—a problem generating or integrating internal signals, say—then the predictions that the brain makes about the internal state of the body and the actual signals it receives don't match up. Perhaps, to make sense of this confusion, your brain then attributes bodily signals and the feelings they produce as coming from someplace else. The result is the sense that you are no longer attached to your body or your thoughts, and that the world around you is going on without you being fully part of it.

And once this peculiar feeling sinks in, it appears to be hard to avoid ruminating over it. This, in turn, can lead to our earlier paradox: despite feeling numb about the world around them, people with depersonalization can still feel an overwhelming sense of internal anxiety.

Her month in hospital was the last time that Louise wondered whether she was going mad. During that time she was referred to Medford.

"I went into his office and told him about what was going on. I was so upset. I just felt like I was the only person in the whole world who was feeling this way, but he turned to me and told me that it sounded like depersonalization disorder. I just thought, 'Oh my God, I'm not insane.' It was such a relief to hear that it was a condition, that I wasn't psychotic, and I didn't have a tumor. It made it all a lot easier to deal with."

Some people with depersonalization find certain mood stabilizers can help control their anxiety, but they are not effective in all cases. Others, like Louise, find cognitive behavioral therapy works well. This generally helps those with depersonalization to break out of the vicious cycle of obsessively dwelling on the strangeness of their internal and external worlds, which can exacerbate their symptoms.

Louise was also shown how to separate her depersonalization symptoms from those of anxiety and depression. "Now, when it's at its worst, I handle it in a much calmer way. I tell myself it's all right, it's just a process happening in my brain, don't panic. Nothing is wrong—I'm still me, that's the important thing. Now when it happens, the rational side of my brain is quicker at responding to it, so that I don't get the absolute panic that was there before."

Louise sits back on her barstool and we listen to the rain hitting the garage door for a moment. It is strangely peaceful, despite the noise and the bright colors all around us.

"I'm not saying that I'll never freak out about it again," she says, "but I feel like I'm prepared for it now, so I don't think it'll ever get as bad as it once was. I've got my weapons now."

Suddenly little footsteps pad down the stairs and a semi-naked toddler waddles into the room.

"Thank God, it's not the same with Morgan and Martha," Louise says quickly, and emphatically. "I've read about people with depersonalization completely dissociating from their emotions. I do have that with the people around me, but not with my children." She stares at Morgan. "Never with them. They actually saved my life. If it wasn't for those two, I would never have got through this and come out the other side."

AFTER SAYING MY GOOD-BYES to the children, I head outside and back into the pouring rain. I sit in the car for a while, watching the water slide down the windshield. It has a calming effect on me. A feeling that occurs only because of the seamless integration of my internal and external worlds. We may think with our brains, but as Aristotle argued all those years ago, we really do feel with our hearts.

I find it incredible that our most basic feelings of existence are underpinned by an ability to sense the internal state of our body. And that being good at it can help us in so many ways. I wondered, Was there any means of getting better at it?

It is often stated that meditation can help increase awareness of our internal body, but there is little scientific evidence to back this up. In fact, when Sahib Khalsa at the University of Iowa tested a group of experienced meditators who practiced either Tibetan Buddhism or Kundalini yoga, he found they were no better at detecting their heartbeats than nonmeditating subjects.[10]

Many other experimental attempts to manipulate interoception have also proved ineffective. It seemed, for a while, that our interoceptive awareness was robust and unchangeable. However, in 2013, Vivien Ainley and her colleagues at Royal Holloway, University of London, showed that the solution may have been staring us right in the face.[11] Her team asked forty-five participants to count their heartbeat while staring at a photo of themselves or at six words that described themselves, such as their first name, their hometown and the name of their best friend. The participants were significantly better at this task when looking at their own photo or staring at these words than when they looked at a picture of someone else or at six random words. It's not yet clear why this works, but the team suggests that focusing on self-referential images and text may enhance the accuracy of interoception by shifting the brain's attention from the outside world to the inside world via the insula.

It has clinical implications: it might help not only people with depersonalization, but also people with more common disorders such as anorexia and depression, both of which have been tentatively linked to low levels of interoceptive awareness.

Whether this kind of training will lead to persistent levels of increased interoception is yet to be investigated. But when we're surrounded by a growing industry of brain-training apps and smart drugs that promise to give you a competitive edge in this world, I like the idea that we might be able to make better decisions, improve our attention and become more empathetic— simply by looking in the mirror.

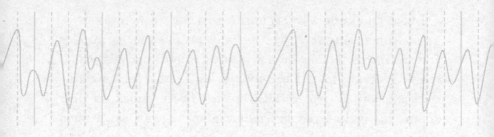

GRAHAM

Waking Up Dead

Turning off the main road and into a maze of one-way streets, I eventually find the entrance I am looking for. I pull up, climb out of the car and stand, watching an old man in a white baseball cap furiously squirt weedkiller onto a patio tile. He holds his back as he straightens up to take aim from another angle.

When he turns to look at me, I quickly walk away, embarrassed to be caught staring. I'm surrounded by aisles of mobile homes propped up on bricks—their yellow, blue and brown cladding faded, a casualty of harsh British winters. Today, however, the sky is blue and I can hear the squawk of sea gulls hovering nearby. Following an unpaved path farther into the estate, I finally spot my destination in the distance. Outside a tiny brown home a man is standing, hands in pockets, clearly waiting for somebody. His face is turned away from me and I slow down, momentarily delaying our introduction.

Suddenly, the man turns and looks in my direction. "Helen?" I smile nervously and nod my hello.

I have no idea how to begin this particular conversation.

Graham is fifty-seven but looks older. His face is freckled and weathered, he has several days of stubble and his hair is receding evenly toward the top of his head. He is wearing tracksuit bottoms and a thick hooded fleece pulled up tight around his neck. I can see his pride and joy, an old maroon Jaguar, parked prominently on his front lawn and know that somewhere across the estate are both his ex-wives, one of whom he still cares for deeply.

I follow him into his tiny home. The porch smells of smoke, and odd bits of carpet line the floor. He leads me through a miniature hallway and points to a faded leather couch.

"Make yourself at home." He has a surprisingly soft West Country accent.

"Right, thanks."

I sit down, and try to muster some sort of tact. "So," I say, as he joins me in the living room, "you used to think you were dead."

If there was ever a brain that deserves to be in these pages, it is that which makes a person believe they are dead. I first came across the condition in 2011, during an interview with Vilayanur Ramachandran, whom *Time* magazine had just placed on its list of one hundred most influential people in the world.

We were both in San Diego to attend the Society for Neuroscience's annual meeting, one of the biggest science conferences in the world, and I had been given the rare opportunity of a face-to-face interview.

Thankful that he had remembered to turn up—he has a notoriously terrible memory—I quickly whisked him out of the press room and into a small adjoining corridor. As we walked, he turned to me. "You know, there are these patients who believe they are dead. They say they can smell rotting flesh but that there's no point in suicide, because why bother when they're already dead?"

It was his idea of small talk. I looked at him in surprise.

"Yeah," he said, his eyes twinkling, "it's really spooky."

WHILE HUMANS BEGAN IMAGINING this scenario centuries ago—corpse-like creatures were the mainstay of the Viking afterlife, and draugrs were the undead bodies of Norse mythology—Ramachandran was in fact referring to a clinical delusion of death that is known as Cotard's syndrome, sometimes referred to as walking corpse disorder.

The medical literature rarely mentions Cotard's, but when it does, it always refers to the French neurologist Jules Cotard as the father of the condition, the person who first described it during the 1880s and by whose name it later became known.

The young Cotard was said to have a "serious and reflective" personality.[1] After studying at the Faculté de Médecine in Paris, he became close to the French philosopher Auguste Comte—a friendship thought to have sparked his intense interest in the mind. In 1864, Cotard became an intern at La Salpêtrière, a teaching hospital in Paris, famed for producing some of the world's greatest neurologists. There he became "a passionate student for . . . madness in its multiple forms."[2]

After a short time performing military service during the Franco-Prussian War, Cotard returned home, where he spent several years in a psychiatric clinic, later opening a clinic of his own in Vanves, a small but densely populated suburb of Paris. Having had the opportunity to study psychiatric disorders from all corners of the country, Cotard became particularly intrigued by severe delusions. It was then that he first described patients with "délire des négations," which he characterized as a type of melancholic belief that leaves a person feeling as if certain parts of their body or aspects of their world have died, or in its most extreme form that one does not exist at all. In 1882, he wrote a chapter for the *Archives de Neurologie*, which paints a vivid picture of the condition. "Patients," he said, "have no entrails, no brain, no head, they no longer eat, no longer digest, no longer dress, and in fact, they resolutely refuse food and often retain their faecal matter."[3]

Some, he added, believe it is their intellectual faculties that have died, that they are being transformed into halfwits, are prevented from thinking, are being told nonsense, and on occasion they even consider that their intelligence has been removed. Sometimes the delusion can relate to the outside world, in which "patients imagine that they have no family, no country, that Paris has been destroyed, that the world no longer exists."

No more than one hundred cases of Cotard's syndrome have since been identified. At least five of these, and probably several more, were described by Cotard himself in lectures and papers he produced throughout his career.

One particular patient of his was the rather exotic-sounding "Mademoiselle X." When Mademoiselle X was asked what she was called, she said that she did not have a name. When probed, she claimed she used to be called Catherine, but wished to speak no further about how she had lost her name. She said she had no age and had never had any parents. When Cotard asked Mademoiselle X, and others like her, if they suffered from headaches, stomachaches or any pain in the body whatsoever, they answered simply that they had "no head, no stomach, no body."

Cotard also wrote about Madame C, who claimed that her throat had been removed, that she no longer had a stomach or any blood. Monsieur C (no apparent relation) refused to wear any clothes because his whole body was no more than a large nut. Monsieur A believed that he had no penis, no testicles, in fact "no longer had anything at all."[4]

When I started this book, I thought a lot about what Ramachandran had said. I asked several doctors whether they had heard of the condition. The few who had, said they'd only ever read about it, and that the people who suffered from it had either passed away—in the more traditional sense—or were dotted around the world in psychiatric care homes, having never fully recovered.

Then one day, out of the blue, along came Graham. He had been a patient of Adam Zeman, a neurologist at the University of Exeter. Zeman told me that he had been treating Graham, who had suffered from Cotard's for many years, but was now apparently in "good fettle" and entirely happy to talk to me.

A few weeks passed while we got the go-ahead from Graham's psychiatrist, and then, as promised, Graham's telephone number landed in my inbox. Which is how I came to be sitting on a leather couch, miles from home, listening to a middle-aged man talk calmly about his recent death.

"So, you used to think you were dead."

"That's right," Graham says, lowering himself onto the couch opposite, seemingly quite at ease.

In the 1990s, Graham lived in this same mobile home but led a very different life. He had two children and only one ex-wife. He worked for a company that provides drinking and waste water for a corner of England. He was a contractor, installing water meters. He was going through his second divorce, and over time became severely depressed. He stopped going to work, avoided his friends and rarely left home. One day, Graham ran a bath for himself, and stepped into it, holding a plugged-in hairdryer.

"Was there something that happened that tipped you over the edge?" I ask gently.

"I don't think so. I was just so low, you know. I don't know how I could have got so low. I don't really like to think about it," Graham replies.

It's not clear what happened next. Graham remembers ringing his brother Martin in a panic; Martin phoned for an ambulance. Graham stayed in the hospital for weeks, cared for by doctors who initially diagnosed him with severe depression. But, unbeknown to them, his depression had morphed into something entirely different.

"What happened while you were there?" I ask.

"I just felt like I didn't have anything in my head," Graham says. "I was convinced that I didn't have a brain, that I'd done something to it in the bath. It was blank. Just a blank mind."

. "And that's what you told the doctors?"

"I told them that I didn't have a brain no more."

These feelings persisted, while doctors tried to work out what was wrong. Often, they would attempt to rationalize. "Graham, how can you walk, how can you sit here talking to me, if you don't have a brain?" they'd say. This dilemma puzzled Graham as much as his doctors.

"It's difficult to describe," he says. "It was like my brain was a sponge that couldn't hold any water."

He lists the side effects of death with surprisingly little feeling. "I didn't really have any thoughts—no emotions. I didn't feel anything. I couldn't smell anything either. I lost my sense of taste. Even my favorite cigarettes didn't give me any hit and I'd been smoking those since I was twelve. I gave them up just like that—normally I'd have been climbing the walls. Nothing gave me any pleasure anymore. I didn't even remember what pleasure felt like. I just had this blank mind and I knew—I couldn't say why—I just knew that I didn't have a brain no more."

"And at no point did you think, 'Okay, I know I must have a brain because I am sitting here breathing?'"

"No. I didn't know what to make of it. I didn't know how I could breathe or talk when my brain was dead. I just knew it was."

This was the challenge that faced Graham's doctors. He could speak, breathe and walk, but was unable to incorporate those abilities into a sense of feeling alive. How, then, do you convince a person that they are alive when they claim overwhelming evidence to the contrary? Graham's doctors tried all sorts of pharmacological solutions, prescribing him antipsychotics and antidepressants, but nothing worked. A scan showed no problem with the anatomical structures in his brain, and no amount of psychotherapy made any difference.

"It just reinforced what I already knew," says Graham. "I told the doctors that my brain was dead—they may as well have been giving me Smarties."

They had reached an impasse: Graham wasn't going to convince his doctors of his death, nor could he be convinced that he was alive. So they all agreed that he could return home under the watchful eye of a community nurse and his brother.

Graham points to the couch. "I sat right there, just like you are now," he says. "All day. For months. I didn't have anything to think about, didn't want to do anything, say anything, see anyone. Just stared at that wall. Like a vegetable. Somehow my body hadn't realized my brain was dead. But I knew it was. Horrible really, thinking about it now. But that was that."

That was that. I close my eyes and think about this unpleasant possibility for a moment. "How did you cope?" I ask.

"What could I do?" says Graham. "I was dead. I just accepted it."

Although Cotard wrote extensively about patients like Graham, the medical community may have made a mistake in naming the disorder after him. In his book *Mental Disorder in Earlier Britain*,[5] Basil Clarke touches upon the work of Levinus Lemnius, a Dutch physician. He describes some of Lemnius's patients—one of whom sounded very similar to Graham. Was it actually Lemnius who was the first to describe the disorder, centuries earlier?

To find out, I paid a visit to Cambridge University's rare books room. It's a large and almost completely silent room, save for the scratching of the occasional pencil across paper—pens are strictly forbidden here. My book was ready and waiting when I arrived: a tiny leather-bound thing, printed in 1581, entitled *The Touchstone of Complexions* by Levinus Lemnius.[6]

Carefully, I carried the old book to the back of the room and placed it on a plush velvet stand as instructed. I was hoping to find mention of Graham's disorder somewhere within its crinkled old pages.

Lemnius was, by all accounts, a popular writer, having published work on astrology, length of life and occult mysteries. *The Touchstone of Complexions* was a sort of early pop-science account of different diseases and why they occur, and claimed to contain "most easy rules . . . whereby everyone may perfectly try and thoroughly know the exact state, habit, disposition and constitution of his body outwardly: as also the inclinations, affections, motions, and desires of his mind inwardly."[7]

Had Lemnius known that Cotard's existed, he would no doubt have blamed it on an imbalance of the humoral system—such was the accepted medical understanding of his time. Indeed, the four humors—black bile, yellow bile, blood and phlegm—and an exposition on their importance in keeping the human organism in balance are the main subject matter of this tiny leather-clad book.

In the final chapter, I eventually found what I was looking for. By this point, Lemnius had reached the brain and was spending some time on the different types of melancholy, paying special attention to patients who suffer from what he called "a depression of the spirits." He picked out a particularly interesting case study. "A certain gentleman fell into such an agony, and fool's paradise," he wrote, "that he thought himself dead, and was in himself persuaded to be departed out of this life."

The friends and acquaintances of this gentleman tried flattering and scolding him in an attempt to restore him to his former strength, but nothing worked. He rebutted everything they had to say and turned away any food they offered, affirming himself to be dead, "and that a man in his state needed no sustenance or nourishment."

That sounded familiar. Doctors had tried to get Graham to eat and drink, but he'd told them he had no need. He wouldn't have bothered with food at all if he hadn't been forced to consume something each day by his family.

When the gentleman in Lemnius's anecdote refused any help, he ended up at the real death's door, so to speak. At this point, his friends came up with an ingenious plan. They dressed themselves in shrouds—cloth normally wrapped around corpses—and sat at a table in his parlor, spread with several platters of food. Upon seeing his friends, the man demanded to know who they were and what they were doing. They answered that they were all dead.

"What? Do dead men eat and drink?"

"Yes," they replied, "and that shall thou prove true, if you will come and sit with us."[8]

Apparently this peculiar brand of logic persuaded the man to feed himself rather well. Disappointingly, Lemnius makes no mention of whether he eventually recovered.

Back in Graham's mobile home, I tell him of Lemnius's story. It seems to make him sad. He tells me that he owes a lot to his family, particularly his brother Martin.

"He made sure I ate during the day," he says. "He still comes to see me every day to make sure I'm okay. It must have been awful for him to see me like that." (I later asked to speak with Martin, to hear his recollection of Graham's illness, but he declined.)

I ask Graham whether any of his friends knew anything about his condition.

"No, I didn't tell anybody. It's kind of a weird thing to say to someone—'I ain't got a brain.' My mates would have just said, 'We've known that for years!' I didn't understand it myself, I couldn't go around telling everyone I was dead. They'd think I was mad."

On the odd occasion that a case study of Cotard's appears in medical texts, it is often accompanied by disturbing descriptions of the patient's experiences. In one, a lady who believed she was in purgatory, having died but not yet moved on, poured acid all over herself, believing that to be the only way to rid herself of her body. This leads me to ask Graham, why, as the weeks turned into months and then years, he didn't try to commit suicide again.

"I remember trying to work all that out," he says. "I did consider it. It was awful really, but the thing was, I believed if I tried to commit suicide again, put myself under a train or put my head on the line . . . well, it's like I told the nurse, I told her, 'I'm sure my head would still be there, I'd still be able to speak because I'm already dead, so the train can't really kill me.'"

Luckily for Graham, medicine has progressed enormously

since the days of the four humors. A few months after his Cotard's began, Graham was referred to Adam Zeman, the neurologist who had orchestrated our initial introduction. Zeman consulted another neurologist called Steven Laureys at the University of Liège in Belgium, because, as he once told me with a smile, "I knew he liked weird things."

"How could I forget," said Laureys, when I asked him about it. "It's the one and only time my secretary has ever said, 'You have to come and speak to this man because he's telling me he's dead.'"

If there are two people you want on your side when you think you're dead, it's these two. Over the course of his career, Laureys has performed some of the most fascinating experiments on the human mind—with somewhat startling results. His lab does all it can to understand, diagnose and treat people who suffer from disorders of consciousness. Occasionally that means discovering that people previously thought to be in a vegetative state with no awareness are in fact locked in—completely aware of their surroundings but with no way of letting anyone know.

In 2006, Laureys and his colleague Adrian Owen developed a test to check whether someone in an apparent vegetative state could in fact follow orders, by inviting them to think about moving around their house or playing tennis. These two thoughts produce very different patterns of activity in the brain, which the team could identify using brain scans. Their first patient—a twenty-three-year-old woman who had fulfilled all the criteria for vegetative state after a road traffic accident—was able to produce the two patterns of brain activity on request. Later they discovered she was very much aware of her surroundings, despite not being able to move, because she was able to answer their questions by attributing the two different thoughts (thinking about moving around her house or playing tennis) to the words "yes" and "no."[9]

Zeman, on the other hand, has concentrated his career on understanding the more bizarre disorders of consciousness, such as permanent déjà vu, that can occur as a result of epilepsy, or more recently, insomnia-induced transient amnesia, whereby people who are severely sleep deprived—doctors, for example—perform complex activities like resuscitation, then completely forget they have done it. Together, the two neurologists have seen more kinds of consciousness than you might imagine could exist.

It might sound strange to talk about different kinds of consciousness when most people think about either being conscious or not. But as we've seen in the previous two chapters, many aspects of our consciousness can take their leave. Tackling the subject of consciousness is not for the faint-hearted. It is a subject that the world's most brilliant thinkers—psychologists, neuroscientists and philosophers alike—have spent many centuries struggling to explain. Most scientists believe that our consciousness, or sense of self, arises from the behavior of a vast assembly of brain cells that act in concert with the body. We can, in theory, map all this neural activity in complete detail and in doing so explain all our behavior entirely in terms of brain states. We can say, for instance, how the brain functions to produce memory and attention and colors. This is what scientists call the Easy Problem. But even if we understand the brain activity underneath all our behaviors, it still does not solve the Hard Problem—why this brain activity results in our rich experience of colors and sounds, or the way pain feels, or the experience of lust. Our conscious sense of self has stubbornly resisted all attempts to understand and describe its existence.

The neuroscientist Anil Seth says that if we want to understand consciousness, we should aim our sights somewhere in between the Easy and the Hard Problem and investigate how certain properties of consciousness arise using measurable biological mechanisms.

For instance, we can start by trying to pinpoint exactly what distinguishes a conscious brain from an unconscious one. Seth says it doesn't seem to have anything to do with how many neurons are active. We know this because the cerebellum at the back of the brain contains far more brain cells than the cortex but can be completely missing without affecting consciousness at all. In 2014, a twenty-four-year-old woman was admitted to the Chinese PLA General Hospital of Jinan Military Area Command in Shandong Province complaining of dizziness and nausea. She told doctors she'd had problems walking steadily for most of her life, and her mother reported that her speech only became intelligible at the age of six. Doctors scanned her brain and immediately identified the source of the problem—her entire cerebellum was missing.[10]

So if it's not to do with the number of neurons, what else might distinguish the conscious from the unconscious brain? A landmark experiment by Adenauer Casali, at the University of Milan, and his colleagues recently tackled this question by stimulating the brain using short pulses of magnetic stimulation. When they did this to people under anesthetic or who were asleep but not dreaming, it created a wave of activity that flowed a short distance from the point of stimulation. However, when they did the same thing in people who were conscious, the wave traveled much farther over the surface of the cortex. Anil Seth later described this technique as being like banging on the brain and listening to it echo.[11] Casali and his team have begun to use this echo to create what they have fondly termed a consciousness meter—a way of working out whether a human, or indeed any kind of animal, is conscious or not.[12]

We can also pinpoint key brain regions responsible for consciousness. For example, there appears to be a group of regions toward the front and top of the brain, called the frontoparietal network, that are vital for consciousness to arise. This network can be further divided into two. Activity in areas along the outside of the frontal and parietal lobes seems to be correlated with our awareness of things in our external world—the smells, tastes and sounds around us. Activity in the second network, distributed among the inner parts of the two lobes, is correlated with our awareness of our inner self—the perception of our body and our mental imagery, for example. When we are concentrating on our external environment, we see the associated network increase in activity as the other decreases. The opposite holds true when we consider our internal self.

In recent years, scientists have also questioned whether our consciousness needs something akin to an orchestra conductor—something that directs proceedings. One proponent of this idea was Francis Crick, a pioneering neuroscientist who early in his career identified the structure of DNA. Just days before he died in the summer of 2004, Crick was working on a paper with his colleague Christof Koch, at the Allen Institute for Brain Science in Seattle, in which he hypothesised that this conductor would need to integrate information rapidly across distinct regions of the brain and bind together information arriving at different times to make sense of the world. For example, information about the smell and color of a flower, its name and a memory of the date can be bound into one conscious experience of being handed a rose on Valentine's Day.

The pair suggested that the claustrum—a thin, sheet-like structure that connects to several other diverse regions—is perfectly suited to this job. The claustrum, deeply embedded toward the center of the brain, is rarely exposed to scientific investigation. However, in 2014, Mohamad Koubeissi at George Washington University in Washington, DC, and his colleagues were using electrodes to record brain activity in a woman with epilepsy when they realized that one of their electrodes was placed upon the claustrum.

When the team zapped the area with high-frequency electrical impulses, the woman lost consciousness. She stopped reading the text she had been given and stared blankly into space—she was awake but not aware; she didn't respond to auditory or visual commands and her breathing slowed. As soon as the stimulation stopped, she immediately regained consciousness with no memory of the event. The same thing happened every time the area was stimulated during two days of experiments.[13]

It's difficult to say that one area of the brain is more important than another in triggering our conscious experiences. I like to think about it as a car: there are many parts of a car that are needed to make it run. Some are more vital than others—you definitely need petrol and an engine and a key or fob, for instance. Perhaps these parts of the car are like neurons, the frontoparietal network and the claustrum. Without one or the other you don't have any consciousness. But there are several other aspects of a car that make it run properly—windshield wipers, a steering wheel, brakes—which I liken to the bits of the brain that help us have agency over our body, that integrate our internal and external worlds, that help us experience color and sound. When one of these aspects of the car goes wrong we can still drive, but something doesn't feel right.

A waft of stale smoke inside Graham's house reminds me of something he'd said to me in passing. Despite having told me that he'd given up smoking, he'd admitted that he would still occasionally smoke the odd cigarette, regardless of the fact that he didn't get a hit from them.

"Just for something to do, like," he'd said.

Something in Graham's manner, in his recollection of this habit, puzzles me. Why, if you truly think you're dead—you don't eat, you don't drink—would you bother having a cigarette? Unless, of course, you still had some kind of craving. It crosses my mind, probably as it had some of his doctors', that I should be taking everything he was saying with a bigger pinch of salt. Just as this thought occurs to me, Graham reaches down to the bottom of his trousers, rolls up the hems and shows me his legs.

"They all fell out you know," he says.

"What did?" I ask, taken aback.

"My hair. I used to have nice hairy legs."

"What, and now you don't?" I say, staring at his naked ankles.

"Nothing! They all fell out. Just like a plucked chicken."

There is a moment's pause.

"I should probably become a diver," he says, his face breaking into the first smile I have seen that morning.

"What did the doctors say?"

"They couldn't explain it. They couldn't explain any of it. I kept on telling them that I'd fried my brain in the bath. But they just wouldn't listen."

And just like that I believe in him.

Zeman, on the other hand, had been convinced of the legitimacy of Graham's claims the moment he met him. "I believed him, yes, absolutely," he told me when I later admitted that I'd briefly had my doubts.

Laureys had needed more convincing. "He's saying his brain is dead. It's very strange when you interact with him, you think it's impossible that he believes this. Of course you think twice; you think, is he tricking me?"

Both researchers, however, were sure of one thing: they needed another look inside Graham's brain. Something had altered Graham's sense of self, and they wanted to find out what.

INSIDE THE CYCLOTRON RESEARCH CENTER at the University of Liège, Graham was placed into a machine that resembled a giant white doughnut. There his brain was analyzed using positron emission tomography, more commonly known as a PET scan. This type of scan monitors all the metabolic activity in the brain—that's all the different cellular processes that are going on at any one time. You'd expect to see quite a lot of activity in someone who is awake.

"What we saw was shocking," said Laureys.

Metabolic activity across large areas of Graham's brain was so low that it resembled someone who was asleep or in a coma.[14]

"I've never seen anyone who was on his feet, who was interacting with people, with activity that low," Laureys said. "And I've been doing this a very long time. Seeing this pattern in someone who is awake is quite unique to my knowledge."

With what's known about echoes and the amount of activity that should be taking place around the brain of someone who's conscious, it just didn't add up. When Zeman and Laureys wrote a paper about Graham, they entitled it "Brain Dead Yet Mind Alive."

While there was nothing wrong with the structure of Graham's brain, his PET scan had shown something very different. First, his frontoparietal network was low in activity. But there were two other regions of the brain that were also problematic.

The first was what's called the default mode network: a collection of neurons that form part of the frontoparietal network but also include some regions of the temporal lobe. Our default mode network is switched on when we're not concentrating on anything. It's associated with mind-wandering, daydreaming and self-referential thoughts. It allows us to think about ourselves, to recollect our past and plan for our future. This ability to think about the things that are happening to us helps us make sense of the world. For instance, right now I can smell bread because I put my bread maker on a few hours ago. I can hear a strange clicking behind my head because my neighbors are doing DIY. My back is aching because my posture is bad and I've been hunched over my computer for too long. My world is making perfect sense. I have my default mode network to thank for that. Graham's default mode network, however, was barely functioning—which might explain why he had such a reduced sense of self. But why did he come to the conclusion that he was dead?

You might think that believing in your own death, despite the presence of overwhelming evidence to the contrary, would require exceptional effort. But perhaps not. The brain hates to be confused. Just as we have seen many times in the book already, when the brain is faced with conflicting information, it tries very hard to make sense of the new scenario, and generally lands upon the simplest narrative to explain an abnormal experience. It's just like the rubber-hand illusion that we saw in Matar's chapter: when we see a brush stroking a rubber hand and feel that same brush stroke on our own hand, our brain comes to the conclusion that the rubber hand must belong to us.

We can demonstrate just how easily the brain can fool itself in people who have a split brain, which happens when their corpus callosum—the region of tissue that connects the two hemispheres of the brain—is severed. This is usually because it has been surgically removed to treat epileptic seizures. Unfortunately some of our abilities are located in just one side of our brain. As we've seen earlier in this book, our basic language skills are normally controlled by an area in the left hemisphere. Because split-brain patients don't have any nerves connecting the two hemispheres, they cannot pass information back and forth between the two. So if you show them something only in their left visual field—which is processed in the right side of the brain—they won't be able to describe it because the information won't be passed from the right side of the brain to the language centers in the left. Let's say you show their left eye a picture of a snowy field and their right eye a chicken. Then get them to pick two corresponding images. In this classic experiment, the split-brain patient usually picks something like a snow shovel and a chicken's claw. But ask them to describe why they picked each picture and their answer is unusual—they would say something like "I picked the shovel because I could use it to clean out the chicken coop." The language area of the brain has access only to what the right eye has seen—the chicken—and has made up a story for why they have picked a picture of a shovel. Here, then, you can see just how easily one's own brain can tell tales, albeit ones it believes are perfectly true.

Simply put, Graham's conclusion that he was dead was likely the most straightforward narrative to explain his bizarre new experience of the world. Once he had come to this conclusion, though, why did he not dismiss this ludicrous idea out of hand? In order to do so, Graham would have to use the brain systems that allow us to evaluate our beliefs. Various pieces of evidence suggest that these brain systems exist in the right dorsolateral prefrontal cortex—the second area of Graham's brain that was particularly low in activity. As Zeman explained it to me: "How can you rationalize with someone if the part of their brain responsible for rationalizing has become irrational?"

I ASK GRAHAM what he thought of his brain scans when he saw them.

"I didn't think anything," he says. "I'd never seen one before,

I didn't know what it showed, just that it showed I had this thing they called Cotard's."

Whether this label gave him any solace is unclear. While it told him that the doctors understood he had a disorder, it didn't give him any self-awareness or new tools to cope with the problem.

"It didn't change the fact that I thought I was dead," he says. "It was just a word that they used to describe my weird brain."

In the year that followed, Graham spent most of his time either at his mother's house, or sitting staring at the wall in his tiny mobile home. There was only one other place he would visit—the local graveyard. Sometimes, he tells me, he would spend the whole day there.

"I just felt it was where I ought to be, you know?" he says.

He would walk around the graves, trying his best to understand his overwhelming urge to be buried.

"It was the nearest thing to death that I could get. I thought, 'I'm brain-dead anyway so I'm not going to lose out on anything, I might as well stay up here.' I felt like I was at home there."

On more than one occasion Graham would go missing and his worried family would call the police. Each time they found him in the middle of the cemetery, happy to spend the rest of his days in a place dedicated to the dead.

At that very moment, on the other side of Europe, there was someone who could sympathize with Graham's condition. She was a middle-aged lady, let's call her Mary, who had just been rushed into Karolinska University Hospital in Stockholm, screaming.

The doctors and nurses were unable to calm her and she refused to tell them what was wrong. Mary's medical notes showed a history of kidney failure and that she'd recently been treated for shingles with an injection of acyclovir. Her doctors decided it was best to put her on dialysis, which would wash out any toxins that may have built up in her blood and that might be causing her pain. An hour later, Mary had found her voice. She said the reason she was so upset was that she was sure she was dead. The doctors tried to reassure her and continued her dialysis. Two hours later, she said, "I'm not quite sure whether I'm dead anymore but I'm still feeling very strange." After another two hours, the woman told the staff, "I'm pretty sure I'm not dead anymore . . . but my left arm is definitely not mine." Within twenty-four hours, her nihilistic delusions had all but disappeared.[15]

The Swedish pharmacologist Anders Helldén and his colleague Thomas Lindén were intrigued by Mary's experience. Helldén said he had started to notice other cases of transient Cotard's, appearing and then disappearing in several patients with kidney failure. He scoured Swedish medical records and discovered eight people who had been in an identical situation over the past three years. They all had a similar story—some kind of kidney failure and treatment with a drug called acyclovir injected directly into their blood supply. You might recognize the name: acyclovir is a common drug used to treat cold sores.

When the pair re-analyzed samples of blood taken from all of the patients, they discovered high levels of CMMG, a molecule produced when the body breaks down acyclovir. Most patients had also developed very high blood pressure.

I asked Helldén what he made of it all. "We have a feeling that CMMG is causing some kind of constriction of the arteries in the brain," he said. Somehow, the parts of the brain affected by this constriction arouse in patients a transient belief that they are dead.

I ASKED ZEMAN whether we could say for sure that Graham's condition occurred as a result of his electrocution. Although it seemed far too coincidental to be much else, I know that such correlations don't sit well with scientists.

Zeman said that it's impossible to say for sure: "Without stronger evidence—a before-and-after brain scan, for instance—we should be hesitant to say that Graham's suicide attempt caused his delusions."

I wondered whether you might expect to see such strange activity in the brain of others with severe depression. Could Graham's brain be an extreme example of this more common condition? The symptoms of depression are similar in many ways to Graham's—hopelessness, loss of interest in life, lack of movement and a detachment from the rest of the world.

The causes of depression are complex, and are yet to be fully understood, but the most recent evidence suggests that the condition may be a result of a lack of serotonin, which is involved in stabilizing mood, and a lack of glutamate, which causes the finger-like tips of neurons to become shriveled so they can no longer pass messages around the brain. I asked Zeman if he thought Graham could have been on the extreme end of this diagnosis, but he said he wouldn't expect to see Graham's brain changes in depression, even if severe. The pattern of low metabolic activity was much more acute and widespread than classically reported in major depressive disorders.

"Of course, with a single case study, one can never be sure," he said, "but Graham's brain changes were exceptional."

There may not be causal evidence to prove that Graham's electrocution triggered his disorder, but we do know that it's not the first time a jolt to the head has led to Cotard's. In the late eighteenth century, Charles Bonnet, whom we met in Sylvia's chapter, wrote a brief report on one of his patients. He described her as an "honourable old lady of almost 70 years." This woman was in her kitchen preparing a meal when a draft came through the door and hit her on the neck, causing her to experience sudden paralysis on one side of her body, "as if hit by a stroke." For four days she was unable to move or speak. When her speech returned she demanded that her friends should dress her in a shroud and place her in a coffin since she had died. She became agitated when her daughter and her friends tried to persuade her otherwise and scolded them for not offering her this last service. Eventually, they did as she requested and laid her out in a shroud. She tried to make herself look as neat as possible, inspecting the seams and expressing dissatisfaction with the color of the linen. According to Bonnet, this woman slowly recovered, although her delusions returned several times a year.[16]

It was with little fanfare that Graham's delusions also eventually lifted. He can't pinpoint the moment he first became aware that he felt better. Whether it was the right concoction of antidepressants, or just a matter of time, his delusions drifted away three years after they began.

"At some point I just came to thinking this is bloody ridiculous, I've got to have a brain," he tells me. Graham's doctors put his recovery down to a combination of drugs and general brain repair. Graham was taking lithium, imipramine and amisulpride, which all modulate chemicals, including serotonin and dopamine, that are vital for controlling the passage of activity around the brain and in doing so can improve mood and help to treat psychotic behavior.

"Gradually I just felt a bit more like myself," Graham says. "Only sometimes I felt a bit dead—but most of the time it was just me again."

He pauses and takes a sip from a mug, which informs me

that he is the "Best Grandad in the World." He points to a picture on top of a side table and smiles. "Lovely my grandchildren are, good as gold."

"Do you see them often?" I ask.

He seems surprised at the question. His simple, fairly emotionless answers to my questions so far had given the impression of him as being a bit of a loner.

"All the time. I see them every week, go round for lunch on a Sunday and see them all."

"What about other people, do you go out much now?" I ask.

"Not on holiday, I'm a bit old for that, but I go down to the club every week, see my mates."

"Do you still see your ex-wife too?" I say.

"Yeah . . . every week." He quickly adds, "The first one, not the second one." His tone turns wistful. "I'm not sure what went wrong there to be honest. I should never have let her go."

Although I have been in his house all morning, talking about a subject that seems, to me, to be rife for self-examination, I am struggling to understand exactly how Graham feels about his strange experiences. He seems to find it difficult to express his feelings, and is somewhat disengaged from his past. At one point he tells me that he hopes his tale will help other people who find themselves in a similar situation—it's a sweet sentiment, but he doesn't seem to realize quite how unusual his experience was.

"I suppose," he says, when I point this out to him.

I wonder whether this was just Graham. A man of few words. Or whether the lack of engagement is linked to his condition.

"Do you feel any different now?" I ask him. "Did the Cotard's change you in any way?"

"Sometimes I wonder, am I different from who I used to be? I don't know. Some of my mates sometimes say, 'You ain't your normal self today,' and I think, 'Am I not? Who am I? What's different?'"

He pauses again to consider the past, and I'm struck by what seems to be a moment of discomfort, the strongest emotion I'd seen so far over the memories of his condition. "It was just bizarre, you know, how could I have felt so odd?" he says. "Just sounds funny now I talk about it."

I wonder, not for the first time that day, whether Graham would ever truly understand the extent to which his Cotard's had changed his life. The only thing he was sure of was his appetite. It never returned.

"It's the only thing that is left over from the Cotard's," he says. "I used to eat regular meals, now I could take it or leave it. I never feel hungry."

I ask him if that was it—an empty stomach, all that remained of the disorder. He hesitates for a moment before answering.

"You know, sometimes I still get funny thoughts now and then. I'm sitting there sometimes and I'll suddenly feel a little bit dead. Just every now and then it happens, and then it goes away."

OUT OF THE WINDOW, I catch sight of Martin, arriving for his daily rendezvous with his brother and I gather up my stuff to leave. As I make my way back to the car, I spot the old man in the cap. He is back outside, squirting at another hardy weed that had popped up between a crack in the pavement. I wave at him and smile.

I leave the estate and drive home deep in thought. Graham epitomizes everything that is so mysterious about consciousness and our sense of self. Here we have a man who can walk, talk and breathe—things that should capture the essence of being conscious—yet for some time, those basic aspects of life were not enough to create a full sense of existence. It is desperately frustrating that the one thing we are unable to comprehend is our ability to comprehend anything in the first place. Perhaps it is because, as the late philosopher Gilbert Ryle put it, "in searching for the self, one cannot simultaneously be the hunter and the hunted."[17] We will always find it hard to examine our own mind when it is the thing that is doing the examining.

We might never solve this mystery. But I take solace in the fact that disorders like Cotard's offer a little glimmer of hope that this may not be the case. For instance, as a result of Helldén's acyclovir studies, there is now (in theory) a way to turn Cotard's on and off at will. Alone, it won't reveal the wizard behind the curtain, or give us all the answers we seek, but it might just propel us another step forward in the endless voyage toward understanding the most complex mystery of the human brain.

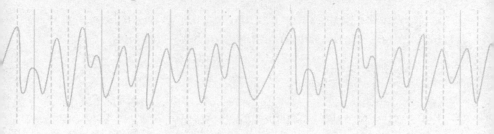

JOEL

Feeling Other People's Pain

Joel Salinas was minding his own business in a lecture theater when he felt someone's hand around his throat. The sensation took him by surprise. It lasted for just a second, before he noticed the lecturer standing at the front of the theater. He had his hand around his throat, and was rubbing it softly. It's times like that, says Joel, when this weird trait can really catch you off guard.

A slight, handsome thirty-two-year-old with brown eyes and a soft voice, Joel grew up in Miami, with Nicaraguan parents who had come to the United States seeking asylum. He lived in a Spanish household, learning English from watching TV. His childhood was relatively straightforward, although there were times when people might have mistaken his precociousness for autism. In fact, unlike many children with autism, he didn't find it difficult to empathize with others or to understand their behavior: he knew only too well what other people were thinking or feeling—because whatever they felt, he felt it too. A scratch of the head, a frown, a slap on the wrist—if he saw it, he felt the same sensations on his own body.

Have you ever watched a painful tackle on the football field and felt a twist in your stomach, or felt sad at the sight of someone upset? If so, you have experienced empathy. It's generally something we cannot avoid; the thoughts, feelings and physical movements of others subtly leak into us, affecting our own thoughts and behavior. It's a fundamental component of human society— one that we are typically unaware of, and that relies on a complex system of mirror-like activity in the brain.

In 1992, Giacomo Rizzolatti, an Italian neurophysiologist at the University of Parma, and his colleagues discovered a group of neurons in the brain that fired in the same way both when a monkey was grasping a peanut and when it was watching a researcher grasp a peanut.[1] These "mirror neurons" were first identified in the premotor cortex—a region responsible for planning and coordinating actions—and later in other areas of the brain, such as those that process our sense of touch.

What makes this mirroring special is that when we see someone performing an action or making a face, we don't just see their actions—our brain, in a sense, feels them too. It is thought that this enables us to internalize another person's actions as if we are the agent in charge of those actions. Over the past twenty years, the existence of mirror neurons in humans has been validated in a number of studies,[2] and they have been described as the driving force behind our great leap forward in human evolution. They are considered by many to be vital to our ability to understand and interpret the actions of others, and to empathize with another person's mental state.

For most of us, this mirroring takes place under the radar. It allows us to empathize with but not literally feel what another person is experiencing because signals from other areas of the brain allow us to distinguish between things that happen to us and things that happen to others. But some people's mirroring mechanism is unusually active, which results in them experiencing tactile sensations and emotions from seeing those same sensations happen to other people. These can be so sudden and so strong that they are sometimes indistinguishable from the real thing.

The condition is called mirror-touch synesthesia. It differs from other kinds of synesthesia that we have seen earlier in that it can have far more visceral results. The first mirror-touch synesthete was discovered by Sarah-Jayne Blakemore, a neuroscientist at University College London. Blakemore was delivering a lecture in which she mentioned anecdotal accounts of people who could feel other people being touched on their own bodies. At the end of the lecture, a puzzled woman in the audience came up to her and said, "Isn't it normal to feel other people's touch?"

In a paper subsequently published in the journal *Brain*, Blakemore reported that she had scanned the woman's brain and showed that her mirror neuron system was much more active than in other people of her age when she observed touch.[3] In the same paper, Blakemore also describes a man who had suffered a stroke that resulted in paralysis and loss of sensation to the left side of his body. When his body was hidden from view, he was unable to feel any kind of tactile sensation. However, as soon as he was able to see his body being touched, he claimed he was able to feel the touch. It was the first sign that tactile stimulation isn't always necessary for us to perceive the sensation of touch—in some conditions vision alone is sufficient.

I was desperate to know what it was like to experience such a condition, so one freezing morning in January, hours before a blizzard buried much of the East Coast of America, I traveled to Boston to meet Joel. Joel had experienced mirror-touch synesthesia for as long as he could remember, but what is most remarkable about his story is his career choice. He is a doctor, which means that he spends the day sharing the sensation of painful injuries, turbulent emotions and even death.

Joel and I are seated in the gigantic lobby of what was once the Charles Street Jail. Famous for housing Malcolm X, the prison was one of Boston's most iconic landmarks. Now transformed into the three-hundred-room Liberty Hotel, the building has retained a lot of its chilling charisma. Each floor has wrought-iron railings, which overlook an opulent central rotunda. Each prison wing now hosts guests who can afford to pay top dollar for such a well-located hotel.

There is something reassuringly familiar about Joel. He is smiley and genuine and an easy conversationalist. Each time I laugh, he laughs too, he is self-deprecating and eloquent and all the things that make you immediately want to be someone's friend. Perhaps I shouldn't be too surprised, since Joel's ability to understand my feelings is far beyond that of a normal stranger. If I were to place my hand on my lap, Joel would feel the sensation of a hand on his own lap. If I bite my lip, he feels a tingle in the same spot. If I were to push a coin along my arm, he'd feel the flatness running along his arm. If I were to poke my leg with a toothpick, he'd feel the pointedness on his leg. He describes the feeling as an imperfect replica of the real thing, "an echo of a sensation." But it's not just other people's touch that Joel can feel, but their emotions, too. When he sees someone looking puzzled, he too feels confused; if someone's angry, the emotion boils up in him as well.

Settled into some comfy sofas in the corner of the lobby, Joel and I order coffee. I ask him to take me back to his childhood.

"Sure," he says. "I'd probably have been described as an emotionally precocious and hypersensitive kid."

He wasn't aware of his mirror-touch synesthesia growing up but, looking back, he believes it definitely had an influence on his behavior. He preferred to hang out with adults rather than children, for a start.

"I think it was because I could take on board their emotional experiences," he says.

"So you absorbed their emotions and enjoyed feeling what adults were feeling?"

"Yeah, I think that must have been it. It wasn't just the happy, sad, scared, angry emotions that you mainly think about as a kid. When I was around adults, I could experience all these other emotions—intrigue, distance, impetuousness. They're words I wouldn't have known at the time, but now I can reflect back and know that I was having that emotion. Experiencing the breadth of their emotional palette was so much richer than feeling what the kids my age were feeling."

In high school Joel tended to talk about other people's emotions—often when they didn't want to discuss them. They would find it really violating, he says.

"Eventually I learned not to do that. I learned to gauge when the time was right to talk about how they were feeling and when to lead them to believe that I didn't have any knowledge of their emotions. It was a bit Clark Kent-ish: I put on the glasses and was just like everyone else."

ALTHOUGH I COULD UNDERSTAND why mirror-touch synesthesia allowed Joel to experience other people's sensations of touch, it was difficult to understand why he felt their emotions too. It wasn't just that he more easily interprets what emotion someone is feeling, he literally feels the same emotion as the people around him.[4] If he doesn't remove himself from the situation or focus his thoughts on something neutral, he can go for hours experiencing an emotion that has no relation to his own state of mind.

Joel explains: "I feel the emotions of others because of the postures that people take on, the facial expressions that they wear, the micro-movements that they do without even being aware of it—those are all felt on my own body."

So although Joel's face doesn't literally move when he sees someone else smile, his brain activity is mirroring the kind of activity that happens when he does smile, which makes him feel like he is smiling, which subsequently triggers a feeling of happiness. Here we've stumbled back upon Antonio Damasio's work—that at the heart of our emotions are bodily sensations.

"I wear what you wear on my body," says Joel. "And that in turn sends a message to my brain that this is the experience I am having. So if someone looks angry, then my brain will feel those movements as if they are happening to my own face and say, 'You're angry.'"

Joel discovered this extraordinary ability when he was in his early twenties. It actually began with the revelation that he had grapheme-color synesthesia. He was in India on a medical school trip when his group started talking about meditation and his friend Elliot mentioned that there were people who can see letters and numbers as having colors and that they apparently had an easier time getting into a meditative state.

"I thought to myself, 'Why did he mention that?'" says Joel, taking a sip of his coffee. "Like, why would that be a noteworthy thing to mention? In my mind, that's just being a person, just being human."

He took Elliot aside and asked him whether seeing numbers with color was normal.

Elliot broke it to him gently. "No," he said. "Not at all."

"And that was my first insight into the fact that I had synesthesia," says Joel.

It was only when he was participating in a trial of synesthetes at Vilayanur Ramachandran's lab in California that he discovered that anything else was amiss. The researchers had asked him all sorts of questions before moving on to talk about his mirror-touch synesthesia—assuming that he knew what it was and that he had it.

"I had this moment of 'Oh, right, so not everyone has this either?' It was genuinely like this light dawning on me."

He says he now has a humbling sense of uncertainty about his perceptions.

"Now I share my experiences a lot more and just have to take a hit when someone says, "Yeah, everyone has that, that's not unique." It's quite comforting when I check about the way I perceive the world and somebody says, 'Yeah that's normal!'"

As well as grapheme-color synesthesia, Joel also experiences the perception of numbers when he looks at people. Not only that, but each of those numbers has a distinct personality.

"So do the personalities of the numbers represent the personalities of the people?" I ask, when he brings this up.

"I haven't tested it objectively to be sure," he says, "but my hunch is that it's quite accurate."

I'm intrigued. I thought I'd heard of every kind of synesthesia, but this one was new to me.

"So how many numbers and personalities are there?" I ask.

"Each number is a small collection of personality traits, almost like a person, so when I meet someone they might have aspects of lots of numbers."

I'm confused, which he immediately senses.

"Look, I can show you," he says. "For you, the most prominent number that I see is the number eight, then there's a couple of ones, with a few zeros. There's also a bit of nine in the background."

"Okay, so what kind of personalities do those numbers have?" I ask.

Joel smiles. "Well, it's hard to describe my subjective perception with the full richness it deserves, partly because it's difficult and partly because I have this scientist in me that's also kind of rolling its eyes. But here goes. So the numbers all have colors. The eight is this bright yellow, a vibrant banana yellow. One is this butter yellow and zero is one of my favorite numbers—it's a clear, brilliant white, and has like a hint of iridescence to it."

"So I'm iridescent and translucent," I laugh, thinking back to Rubén's description of my aura.

"Eight is like a hardworking, strong, earnest person with true intentions. Number one is also true, but can have a competitive edge."

My family would definitely say that bit was spot on, I think. "I feel like I'm getting my palm read, Joel."

"Well, yeah, I mean this is the basis of cold reading, right?" he says, continuing with his description. "Nine is a very black number—it's what I associate with an executive, high-powered person. Someone who is firm in their actions and can command a room if they want to. And the zero that I see has this Zen, calm neutrality to it."

Needless to say, I find it pretty disconcerting to have someone sum up my personality, having known me for less than fifteen minutes. But to some extent it is what we all do. We judge the people we meet instantaneously, sum them up in our minds, box them into categories of who we think they are. Most of us just don't have it all painted out quite so clearly; more often these judgments are a murky instinct that we would perhaps refer to as gut feeling.

"Do people's numbers change?" I ask Joel.

"It's kind of like focusing an image," he says. "The more data I gather about a person, the more the image will refine and numbers will be added in different places and sizes and then when I know someone well enough, it becomes a landscape made up of the colors of the numbers it represents. One friend, for example, is this turquoise lagoon in a gray crater, because he has a lot of aspects of seven, four, some sixes, rare zeros, but a lot of seven, which makes up most of the water."

"What traits does seven have?" I ask.

"Seven is an endearing kind of weirdness. Like someone with a slight quirkiness, you know? But you just gotta love 'em!"

"Do you see numbers when you look at your own reflection?"

"Yeah, it's not exactly a number. It's like an experience of shining a flashlight at a mirror, kind of like a bright light without any additional information to guide me. The closest number I could pin it to is zero. I'd like to say I see fours, but it's probably more the desire to see four because it has characteristics that I want to emulate in myself. It's like a calm, soothing, friendly number. A gentle storm before the rain rolls in."

"Do you think the colors and numbers you see in other people influence your opinion of that person?" I'm thinking about the trouble that Rubén sometimes has separating the two.

"Yeah. When I was younger, I'd have reactions to people that were associated with these perceptions and it would turn me off or on to them. But as I gained a lot more awareness of the process, I've also been able to gain more objective distance and now ask myself, 'Does this make sense? Is this my implicit bias? Is my discomfort with this person because they have a lot of five? Should I be giving the three a little more benefit of the doubt?'"

"Do you ever ignore the numbers completely?"

"Sometimes, but I find that I actually tend to get burned

more for ignoring the associations. It's almost like looking under the hood of intuition. And to ignore it, you'd also be ignoring your instinct."

I look around the lobby. There are people milling around, heading in and out of the hotel and sitting drinking coffee, working on their laptops. I wonder just how far Joel's sensations actually reach. Can he feel what these people are feeling? I ask Joel to describe a snapshot of his world right here and now. He snatches a glance at the three people seated nearby on a long sofa. "I can feel the flatness of the phone that woman had against her cheek," he says. "And then the man next to her sort of did this shrug where his head went into his neck." Joel dips his chin into his neck. "I felt his double chins on my chin. And then did you see that woman rushing past? I felt her hair on the back of my neck, the way it was brushing back and forth."

I was about to ask Joel how on earth he gets anything done when he feels so much of the world around him, when he surprises me yet again.

"You see that there," he says suddenly, pointing to a long, thin vase sitting in the middle of the table between us. "I feel that on my body too."

"You feel inanimate objects?"

"Yeah, if I look at it I can feel the length of its neck in my neck." He stretches his head upward. "It feels like my neck is elongated, holding my head up high.

"Sometimes, I'll get irritated and angry and I'll look around and notice an object that was in my peripheral vision that resembles an angry face and be like, 'Oh right, it's because of that.'"

We all begin mirroring others at an early age. Poke your tongue out at a newborn baby and you'll see what I mean—they'll poke their tongue right back at you. We inadvertently mimic people in others ways too. The former UK prime minister Tony Blair was famous for being a linguistic chameleon, changing his accent to suit his audience. In fact we all have a tendency to imitate other people's accents, as well as their facial expressions, body language and mannerisms. Some studies suggest that people respond more warmly to those who subtly imitate their movements. This unconscious mimicry acts as a kind of social glue: if our body language is alike, then our mental states must also be similar. A word of warning, though: consciously trying to imitate someone to make them like you can divide your attention and you'll likely end up with the opposite result.

Despite all this mimicry, few of us have to make a serious effort to distinguish what happens to other people from what happens to ourselves. Yet Joel's brain seemed to have difficulties separating the two. To find out why, I paid a visit to Michael Banissy, a neuroscientist whose office in Goldsmiths College was just a few miles from my home in southeast London. His lab works closely with all kinds of people who have difficulties with social perception, trying to understand how these abilities vary between us. He has scanned the brains of several people with mirror-touch synesthesia, including Joel, and thinks he knows what drives their strange perceptions.

Brain scans show that the mirror neurons in mirror-touch synesthetes are overactive when they see other people being touched. It's likely that there is some kind of threshold that you have to reach to become consciously aware of a tactile sensation, and that people with mirror-touch synesthesia cross over this threshold when merely looking at other people.[5]

But if our mirror systems are active in the same way when we see someone touched and when we feel touch ourselves, why don't we all go around sensing other people's touch on our own bodies? One of the reasons is that when you see someone touched, tactile receptors in your skin aren't stimulated, so they send messages to the brain saying, "I'm not being touched." This signal vetoes some of the activity of the mirror neurons. Sometimes amputees can feel touch on an area of their missing limb when they see others being touched in the same place. They're not getting any of the normal veto signals from the skin since there's nothing there to send them. But what's allowing Joel's mirror systems to run amok?

To answer this question, Banissy's team ignored the mirror system entirely and went in search of strange activity in other areas of the brain. What they found was quite remarkable. Mirror-touch synesthetes seem to have less brain matter in their temporoparietal junction—an area that is said to help us distinguish the self from other.

"It's like they have this fuzzy boundary between themselves and other people," said Banissy. To test this idea further, he got eight mirror-touch synesthetes to take part in a game in which they had to raise one or two fingers while watching others do the same. They had much more difficulty completing the task when the person they were looking at was raising a different number of fingers than they were told to lift.[6]

"It was like their brain had some difficulty inhibiting the idea that this other person wasn't them," he said.

Left to its own devices, Joel's brain activity seems to mirror the world without restraint, crossing the threshold past which other people's perceptions become his own.

LATER THAT EVENING, Joel and I brave the cold again to meet for dinner at Clink, a restaurant at the Liberty Hotel. Joel rushes up, just a few minutes late. He explains that on his way over he'd had to work hard on getting rid of an emotion he'd experienced ten minutes earlier from a colleague. It was the worst kind of emotion, he says: passive aggressiveness.

"That emotion, that kind of malice, really clearly stands out for me," he explains, as we sit down at our table. "I have to take a step back and remove myself from the moment because I get this knot in my throat. It's so vivid I'm like 'Urgh, God, that's painful!' and I don't want it to become confrontational. Mostly my emotions change pretty quickly but then there are occasions like these that it'll stick and I'll take on this irritability that I really have to work hard to get rid of."

Joel also dislikes it when people deliberately attempt to hide their emotions from him. "If I can see what your actual emotion is below this façade that you're playing then it really stands out to me. I think it's really amplified in me."

"That must happen a lot in the hospital, no?"

"Yeah, it can. Sometimes a patient will say they are fine, and I know they're not because I can really feel some strong negative emotion. Like I know they're about to cry because I'm about to cry. But most of the time it definitely helps me. I can't say that I'm feeling exactly the way they are feeling, but I can feel their discomfort and their distress, or if they're scared or confused or feeling better. Sometimes it's hard to know where the mirror-touch starts and where it's just normal human empathy kicking in."

Within the hospital environment, it's difficult to understand how Joel keeps his cool. If a person is in pain, coughing and heaving, he can feel his own lungs tightening. When they are intubated, he feels the tension in his own vocal cords as the tube is placed down their throat. When he injects a needle into someone's spine, he feels the sensation of a needle slowly sliding into his own lower back.

It's not just his patients' physical distress that he feels, but also the emotional anxieties from their family and the nurses. Directing his attention elsewhere is the (very successful) trick he has learned to regulate this sea of emotions.

"I try to concentrate on the calmest person in the room, or just stare at my sleeve or something," he says. Although sometimes in the middle of a busy emergency room his synesthesia is unavoidable. "There was this one time in medical school when I saw someone's amputated arm. I had these vivid physical sensations of my own arm having been ripped apart. It was really difficult. I think it was so vivid because it was something I had never seen before—novelty seems to have more effect on me than things I see a lot."

Of course, sometimes this intense empathy comes in handy for diagnosing a patient or finding out what's going on underneath the surface. The physical sensations that he experiences from others and the heightened attention to micro-movements has made him a much better observer, he thinks. "I can pick up on subtle twitches and movements of the eyes and mouth that others might not pick up on. It can help lead me to a quicker diagnosis or understand a bit more about the complexity behind what's going on."

"Are there moments in which you particularly utilize your hyper-empathy?" I ask.

"I do when I see patients in really tough situations. It's something most patients are starving for—to feel that there's a connection between them and the person who is caring for them. I also rely on it when you have to tell people they have a terminal diagnosis, like Alzheimer's. It's never an easy conversation, but it's made harder by the fact that the person affected may have a bit of insight into their disease to know something's wrong but not enough brain capacity to really understand what's happening. So I use it then to try to get in tune with as much of the person beneath the disease as possible."

He says it's a bit like windows on a computer desktop. "I can choose to maximize certain windows to really hone in on a feeling and make it all the more vivid, but there's always this wash of emotions being processed underneath that affects everything that I do."

"So can you ever turn it off completely?" I ask. "Ignore other people's emotions around you?"

"No, there's always this white noise, a haze of things going on. I'm almost a fool to believe that any of the emotions I'm feeling are completely my own."

It suddenly occurs to me that Joel must have seen plenty of people die. I ask him what he feels in those moments.

"To put it simply," he says, "it's as if I'm dying as well. There's this very powerful moment just before death of letting go. It's not so much the presence of a feeling but the absence of one. It's kind of like when you're in a room with an air-conditioning unit in the background and suddenly it just shuts off. There's this disquieting stillness."

The first time Joel ever saw someone die, it was unexpected—a man lying on a bed near to him, waiting to be taken elsewhere in the hospital. Joel's body mirrored the man's. Suddenly, he felt his own breathing slowing. It wasn't Joel imagining what death must be like, but his body physically imitating the process. "I needed to start having to be more voluntary about my own breathing, otherwise I felt like it would just stop as well."

On hearing this, I wonder why Joel was drawn to this career in the first place. In some ways it seemed a natural fit, in others a complete nightmare.

He says one inspiration to becoming a doctor came about after a stint as a medical assistant with his uncle in a rural Louisiana clinic. "I saw how important it was to the community, and I'd always known that I wanted to help others. I thought about all the things that make me happy and give my life energy and purpose, and it ended up being a collage of things that were compatible with medicine."

He says he watches horror movies and psychological thrillers at home to help him cope with the unexpected at work.

"I know it sounds strange and I know other mirror-touch synesthetes would be burdened by it, but I see it as part of my education. It helps me learn more about others, and manage crises. What good is a physician who shuts down at the sight of blood or violence? The more novel and surprising an experience is, the more vivid the synesthetic experience is for me, so I expose myself to it so that it's less novel when I see it in real life."

"Do you think you'd do the same if you weren't a doctor?"

"Yeah, I think I'd still see it as part of the development of my character. It's almost like a way of experiencing the world and living a full life. I wouldn't want to short-change myself."

IT'S NOT JUST JOEL who can overdose on other's feelings—all of us can risk becoming infected with other people's pain. It's something known as emotional contagion. Our emotions can spread like a virus—with some truly awful consequences.

Our capacity to understand others' feelings through empathy is crucial for successful social interactions—it was this empathy that may have given us a giant kick in our evolution as a social, collaborative and moral species—yet empathize too much and you can actually make yourself sick. Nurses, in particular, are at high risk of this kind of emotional burnout. The consequences are bad for their health: they experience increased levels of anxiety and stress, but also anger, aggression and overall lower levels of empathy.

You might think you're immune to such social contagion but several experiments suggest otherwise. In 2014, researchers played with our emotions by tweaking Facebook's algorithms so that certain people were presented with more negative or positive posts. They showed that these people became more negative or positive themselves as a result.[7] Experiments on Twitter users have shown a similar effect.

While some people are naturally better at empathizing than others, it is possible to change your natural state. In 2013, Christian Keysers from the Netherlands Institute for Neuroscience and his colleagues tested this theory in twenty-two male offenders diagnosed with psychopathy and who were thought to have low levels of empathy. He showed his volunteers videos representing people in love, in pain or experiencing social exclusion, while scanning their brains. The results showed that people with psychopathy had much lower activity in areas of the brain responsible for empathy compared with a control group that had no history of psychopathy. Activity was particularly low in the insula, which as we've seen in previous chapters is vital for coordinating signals from the brain and body. However, when Keysers's team instructed their subjects to try consciously to empathize with the people in the pictures, the brain scans of those with psychopathy matched those of the healthy control group.[8] It suggests we may all have a whole spectrum of empathy within us that we can choose to ignore or ignite.

So how do we empathize without burning out? A series of studies, many by Tania Singer at the Max Planck Institute for Human Cognitive and Brain Sciences, in Leipzig, Germany, suggests we should transform empathy into compassion.[9] We often use these words interchangeably, but they mean different things. "Compassion" can be described as having caring thoughts for another person—for instance, when a mother reaches out to a screaming child. "Empathy" is putting yourself in another person's shoes and vicariously experiencing their emotions. When Buddhist monks are asked to engage in a form of compassionate meditation while listening to distressing sounds such as a woman screaming, areas of the brain involved in empathy, such as the insula, decrease in activity. When people who are not trained in compassionate meditation are asked to listen to the woman screaming, pain networks light up in the brain.

Singer wondered whether short-term compassion training could help people to think more like monks. After just a few days of lessons, their brains began to look more like the meditating monks' in response to hearing other people in pain. They are still able to feel *for* them, but no longer feel *with* them, and early results suggest this leads to an overall sense of increased well-being.

If you'd like to try it for yourself, compassion training simply involves spending time thinking about extending warmth and caring feelings—like those you would usually experience toward a much-loved person—to everyone around you. By concentrating on compassion, rather than empathy, you could protect yourself from emotional burnout.

Over dinner, Joel tells me about a couple of times that he has been a patient himself. One time was after a terrible car accident, in which his car rolled and left him in intensive care with lacerations and a cervical collar around his neck. Now, whenever he sees someone his age with a cervical collar, his sensations are at their most vivid because he knows exactly what they feel like. His second experience of being in a hospital was even more dramatic. It was 2005, and Joel was in Haiti working with the local government, providing medical services to hard-to-reach areas of the country. During the trip, Joel developed a sudden headache. "It was different to a migraine—it was really specific to the right side of my head," he says.

Luckily, there was a neurosurgeon on the trip.

"What does it mean if you develop a sudden headache?" Joel asked him, casually.

The surgeon joked, "Oh, it usually means you're going to die."

"I was like, 'Oh right, 'cause I have that.'"

When he got back to Boston, the neurosurgeon had two of his assistants give Joel a full examination. They found what looked like a tumor embedded above the brain, eating away at the skull. It wasn't clear whether it was attached to the brain or not. It needed to be removed.

In the operating room, a surgeon peeled back Joel's skull to find a pulsating mass of blood vessels. They pulled it out, cauterised the bleeding and filled in his skull with bone cement. Thankfully, the tumor wasn't malignant. When Joel came around from the anesthetic, the first thing he did was search for a letter. He wanted to know whether his surgery had affected his synesthesia.

"I looked for a letter to see if it still had a color or not—I was really thankful that it did."

Whether Joel's brain tumor had anything to do with triggering his mirror-touch synesthesia is unclear. But the tumor was near to his temporoparietal junction. If he'd had a bundle of abnormal blood vessels growing there since birth then there's a chance that he might have had greater vascular supply to that part of his brain, making it develop abnormally and perhaps resulting in this unusual blurring of his self and others.

As we eat, Joel tells me that he has been having an especially difficult week. He's been leading a Tourette's clinic. One of his patients has a self-mutilating tic, which involves biting the side of his cheek, pushing on his face and grinding his teeth.

"It is a massive challenge for me," he says. "Most of these tics are very surprising so it's a perfect recipe for me to feel them. I have to be really conscientious not to start copying the tics.

Every now and then I have to take some time out and stare at the computer screen or the floor and remove myself from the situation."

A few days ago, Joel's patient was continuously pushing on his face with his knuckles, creating cuts so damaging that he'd had to have surgery on his mouth. It was a particularly tough session because there was a lot going on in the room.

"Every time the guy would tic, I'd feel like there was a fist mashing my face," says Joel. "I could feel my lip up against my teeth almost like it was being cut."

Then there was a moment where he was completely caught off guard. "The patient pushed on his face and ground his teeth in this way that was so loud, and as he did so I felt this vibration across my face that was so extreme. It really exited the realm of an internal perception and became this very real experience."

I wonder what Joel does to get away from it all—to relax. He says he exercises a lot. I'm amazed to hear that even here his mirror-touch synesthesia helps him out. "I tend to learn new physical skills easier than most people," he says. If he's watching a tennis instructor demonstrate a serve, for example, he can feel the movements in his own body, meaning that when he repeats the movement for himself he can tell if it matched or not, and if not, where he went wrong.

He runs every day that he can. He often watches Japanese manga while he's on the treadmill because there's a lot of running in it. "If I'm running, and they're running, there's no mismatch, and for that short time everything makes sense in the world."

SPEND ENOUGH TIME WITH JOEL and it's hard to ignore the strange sensation that he knows you like a best friend. He finishes your sentences and immediately senses when you're confused or troubled. But sometimes this can make relationships difficult. Over the past year, he's been going through a divorce—a difficult situation at the best of times, but if you're a mirror-touch synesthete it's all the more complex. That's because in an argument Joel takes on board the emotions of the other person. And when you're trying to iron out your difficulties, too much empathy for another person's feelings makes it difficult to keep your own feelings straight.

His ex-husband lives in Seattle and at the worst point of the divorce they talked by FaceTime. It helped, says Joel, to have an image of his own face in the corner of the screen in an argument.

"The minute I'd feel that I was putting myself way too far into his perspective on things, I would look at myself and get back to how I really felt."

"It sounds complicated."

"Yeah, it was. The minute that I did something, it was affecting him, which was then affecting me and it turns into this really turbulent spiral."

I wonder how Joel's life would have turned out had he not been so bright, so willing to understand his strange brain. He says that if he didn't have the intelligence to understand and manage these experiences, his world could easily have come crashing down around him. "All these experiences could be really anxiety provoking," he says. "My world could be ruled by them. And that would be interpreted by the medical profession as schizophrenia, psychosis or mania of some kind."

Someone suddenly laughs loudly on one side of us. I wonder if it had made Joel slightly happier for a second. But then the couple on the other side were looking serious, deep in conversation—perhaps theirs was the emotion he was feeling.

Before I met Joel in person, it was hard not to feel as though I was missing out on some grand superpower. How often do we complain that someone is hard to read, or that we wish we knew what someone was feeling? But when it comes down to it, would we really want to know? It must be exhausting, rebounding from emotion to emotion throughout the day.

"Yeah, it can be," he says. "The more I lose my reserves, the harder it becomes for me to manage other people's emotions. But it can be such a lovely thing to have. If I'm feeling upset, I can reflect on it and ask myself, 'Is this feeling of irritation springing forth from me or is it me reflecting the experience of someone else?' And if the latter, I'm able to remove myself from that emotion, rationalize it, extinguish it and then address what is causing the irritation in the other person."

It's like learning how to surf, he says. "You've got this whole emotional ecosystem going on below you that's constantly moving but if you can understand its movements, and embrace it, you can move along with it and enjoy yourself. If there's a big wave—whether it's negative or positive—you can enjoy that wave."

"Do you ever just hang around with someone because you can see they're happy and want a bit of that emotion?" I ask.

He laughs. "Yeah, for sure! I deliberately smile at people so they smile back at me so I can get a hit from them."

"A hit? Like a boost that makes you feel better than you could alone?"

"Yeah, exactly. I love watching people hug. It's very warm and very comforting. And I tend to be very affectionate and congenial, and a lot of that comes from the fact that I genuinely want people to feel good, but you know, also it's nice when they don't have negative emotions because I don't have them either. When they're positive I get some of that as well. That sounds both really selfless and selfish—I guess I'm actually just a selflessly selfish guy!"

Toward the end of our meal Joel points to the painting above my head. It is full of meaningless black and brown and white swirls but he says it looks completely different to him because the swirls resemble letters and numbers, which pop with color. I ask what else he was experiencing from around the room. I am expecting him to talk about the people sitting next to us, but instead he immediately points out that he can feel a hand on the back of his neck—it's my hand, brushing away my hair. I smile and quickly place my hand in my lap. He says he can feel the bite of where I'd just bitten my lip. "And now I can feel the touch on the side of my face where you just touched your cheek, and now the slight tensing of the corner of your mouth and now the squinting of your eyes and the—"

"Stop!"

I am suddenly hyper-aware of every single movement that two seconds before I hadn't even known I was making. In that moment, I catch a glimpse of how much of an assault on the senses Joel's life really is.

"And that's why I tend not to talk about it with a lot of people," he says quietly. "It makes them feel really awkward."

"Yeah. It's kind of difficult to concentrate when I know you're feeling everything that I'm feeling."

There is a brief moment of silence.

"How close do you think your emotions match other people's?" I ask.

"Sometimes I feel like it's pretty accurate, which is when it sort of feels like there's some kind of mystical quality to it." He laughs. "The scientist in me is throwing a fit at that phrase. But most of the time it's an imperfect perception of what you're feeling—I can't take a sci-fi quantum leap into your body. It's almost an insult to you, for me to ever assume that I could feel your pain and your emotions exactly as if they were my own. For me to say I know exactly what you're feeling is kind of rude, and . . . violating."

It crosses my mind that he could be playing down his abilities to make me feel more comfortable. Perhaps he doesn't want to reveal just how closely matched our feelings are. I sit quietly and bite my lip. I immediately wish I hadn't. Then immediately wish I hadn't thought about not biting my lip because it's made me frown. And then I brush my hair aside again. I am conscious of everything I am doing. Suddenly I feel a yawn coming on. I came to Boston from London via a week working in Texas and Phoenix, and after all the time changes and the traveling, I am jet-lagged and exhausted. But as I muffle the yawn I realize there is probably no point—if Joel is feeling what I am feeling, perhaps he already knows I am tired, perhaps he knows I am trying to stifle a yawn, perhaps he thinks I am bored. How should I make it clear with my facial expressions that I am genuinely fascinated by our conversation, just tired from traveling? How does my face normally look when I'm fascinated? I am lost down a rabbit hole of self-analysis and I have completely missed what Joel has just said.

And it is no use trying to act like I haven't.

Over a shared dessert of cheese and crackers, Joel and I discuss something Ramachandran told me many years ago about how humbling it was to learn that the only thing that separates two people is a layer of skin.

"Mirror neurons make us all alike," Ramachandran had said. "They're acting in the same way whether you or I make the action. If you remove my skin, I dissolve into you."

While Joel may feel this in the extreme, like many of the extraordinary people I have met while writing this book, it is not a characteristic unique to his brain, but an extreme example of an ability we all possess.

Joel agrees. "There's this constant hum of other people's experiences that is going on all the time," he says. "I might feel them more strongly than other people, but it's something that affects us all."

It's a delightful concept and one that we would do well to remember. That our brain does not exist in isolation. We discovered that it relies on our bodies earlier in this book, but its reach stretches farther still. It extends beyond the boundaries of our skull and enters the bodies of those around us. In that way, we are all connected with one another. When we smile at someone, we leave a tiny imprint on that person's brain. Somewhere, deep within their motor cortex, their brain is smiling back.

Conclusion

NOTHING IS UNTHINKABLE

First thing in the morning in the midst of spring, the south coast of Norway is awash with the heavenly scent of sea salt and pine. The main highway weaves in and out of the jagged fjords, which are lined with green and orange trees, giving you the occasional glimpse of the ice-blue sea between.

Four hours' drive from the sprawling metropolis of Oslo city center, down toward the southernmost tip of the country, is a small but beautiful coastal city called Arendal. Looking out on lots of tiny islands and considered to have the world's most beautiful sea approach, the city itself is packed with wonky wooden houses, cobbled streets and colorful bars.

But I hadn't traveled seven hundred miles for sightseeing. I was there to visit a small office-supply company called Østereng & Benestad.

A month earlier, I had been packing up my notes, filing them away in boxes to put in the loft, when out fell the crinkled old Jumping Frenchman paper that had been the source of inspiration for this book.[1] I sat crossed-legged on the floor of my study and reminded myself of the story.

It was 1878, and George Miller Beard had traveled to Moosehead Lake in northern Maine. He had heard about a strange disorder afflicting some of the men who worked in the area. The locals cheerfully referred to them as the Jumping Frenchmen. They were of French-Canadian descent, and spent the winter working as lumberjacks, completely isolated from civilization. It must have been summer when Beard first visited, because he came across his first two Jumpers working in his hotel.

One of them agreed to let Beard perform some experiments on him. Seated in a chair, the young man began cutting his tobacco with a knife. Beard struck him sharply on the shoulder and told him to "Throw it." As quick as a pistol, the man jumped and threw the knife, so hard that it stuck in the beam opposite. Later, Beard shouted at him to "Strike" while standing near another employee. Immediately and without hesitation, he struck his colleague on the cheek. When mildly kicked on the shin or tapped suddenly on the shoulder he would leap and cry out. He knew he was being studied, yet could not help but explode at even minor knocks and taps.

Beard observed another Jumper of just sixteen. The crowd in the hotel, partly for Beard's benefit, teased the teenager so much so that he was constantly on edge. His caution was fully justified. When he was standing close to another Jumper, a stranger yelled, "Strike!" They jumped and struck out simultaneously, hitting each other in the face. These were not, said Beard, "mild or polite little prods," but "severe and painful blows."

During his time at the lake, Beard met many other Jumpers.

Among them was a waiter who would let go of whatever was in his hands whenever anyone shouted, "Drop it!" On one occasion, this resulted in him dropping a plate of baked beans onto the head of a hotel guest.[2]

SO WHAT WAS BEARD'S CONCLUSION? The men, he said, were in their prime. They were strong and extraordinarily healthy from all their physical labor. It did not strike him as a disease; instead he felt that the disorder was some kind of learned affliction resulting from the constant reinforcement of a naturally high startle response.

We all startle—it's a defensive reaction to sudden noises and movements and can sometimes save our life. It's part of our fight or flight response, a reflex that happens automatically without conscious control. It increases our heart rate, directs out attention toward potential danger and shoots out hormones that fuel our subsequent actions. It varies considerably between people. My husband, for instance, jumps at things on the television that aren't remotely startling to me. People with post–traumatic stress disorder can have an overactive startle response, due to associating powerful emotional memories with sudden noises. This puts their brain on high alert, lowering the threshold for a future reaction. It can also be modified by your surroundings: if someone jumps out at you in a friendly game of hide-and-seek, say, you'll startle less than if a stranger jumps out at you in the middle of a dark alley. Beard's Jumping Frenchmen seemed to have been picked out for their naturally high startle response, which was then exaggerated by their friends' and colleagues' constant provocation. This was aided by the fact that their reaction was one of the main forms of entertainment in the isolated woods. The attention they got from jumping was almost always positive: people would laugh, a reaction that reinforces our behavior in any context.

Sitting in my study, I wondered again what had happened to these men and whether the condition still existed. To find out more, I set up an interview with Marie-Hélène Saint-Hilaire, associate professor of Neurology at Boston University School of Medicine. I explained how their story had encouraged me to write this book. I told her that I thought she might have been the last person to meet a Jumper.[3]

"It's interesting, isn't it," she said, "how people seem to be really taken by this condition. I think it's all in the name."

Back in the 1980s, Saint-Hilaire was a medical student in Montreal. One day, her neurology professor asked whether she had ever met a Jumping Frenchman, since she had been brought up in Quebec, an area close to Maine.

"I'd never met anyone like he described," she said. "But since I was about to do some rotation work, which had to be in a different location, I decided to go back to my hometown and ask my grandfather if he knew anyone that jumped."

Her grandfather told her, "Sure, the guy down the street is a Jumper. It's great fun to make him jump—we'd do it all the time when we were little."

Saint-Hilaire and her father, a neurologist, decided to go and speak to the Jumper and film their interaction. "We asked him about other Jumping Frenchmen," said Saint-Hilaire. "He told us about two other guys, as well as his sisters."

All of the Jumpers were either lumberjacks themselves or related to a lumberjack. The men told Saint-Hilaire that they would spend the summer working on the farms and in hotels, and in the winter they'd head into the forest. There they would spend six months without leaving the camp. At the beginning of the season, it was traditional to find out who was a Jumper and to begin startling that person as much as they could.

Saint-Hilaire filmed the men while her father asked about their medical history and performed a neurological exam. I got in touch with the journal that subsequently published a paper that Saint-Hilaire and her dad wrote about the Jumpers. They still had the video—now almost forty years old—in their archives and sent me a copy. It opens with an image of a seventy-seven-year-old ex-lumberjack sitting in a La-Z-Boy chair covered with a thick leopard-print blanket, surrounded by pictures of his wedding day. Saint-Hilaire's father is off to the side, sitting on a stool, inquiring about his days in camp. Suddenly, he makes a noise and lunges toward the old lumberjack, prodding his leg and torso. "Whooaa!" shouts the old man, his legs flying into the air and his arms waving from side to side above his head. Both men laugh and there is giggling in the background from behind the camera.

"My father made them jump while I filmed it," said Saint-Hilaire. "It was very interesting. They described this elaborate reaction that occurred when they were young, but it had become muted with age. We assume because they were no longer startled very often. They're older, away from the environment that started it, so the reactions were still exaggerated but not as marked as they once were."

They no longer repeated commands or carried out automatic actions that were demanded of them, although most still had a defensive reaction.

"My father once screamed at this one older lady to 'Dance!' and she didn't dance, but she did try to punch him."

"Are any of the Jumpers you spoke to still alive?" I asked.

"No, they've all passed away now," she said. "I think the condition ended with them. Soon after that time, life in the camps changed. There was more machinery, more technology— they were no longer as isolated or in need of that kind of entertainment."

And so there the story finished. I packed up my papers and taped shut the lid of the last box, thinking my journey had finally come to an end.

ONE YEAR LATER

I STARED AT THE SCREEN and watched the clip for the third time that morning. It was a YouTube video sent to me by a friend.

"Isn't this like that Jumping Frenchman disorder you talk about?" she'd said.

The video was entitled "The most easily scared guy in the world?"[4] It consisted of a short excerpt from a Norwegian TV broadcast revealing the antics of Basse Andersen and his colleagues at paper company Østereng & Benestad.

Basse was a gray-haired, middle-aged man with a strong Scandinavian jaw, black-rimmed glasses and a wide smile. In the video (which currently has more than three million views) Basse's colleagues ask him to collect a box from the warehouse. Little does he know that the box is in fact covering the head of a person hidden inside a larger box beneath. As Basse lifts the top box, he discovers the head and screams, runs backward and falls over. The interviewer talks to Basse's colleagues, who say he likes being the center of attention, that he finds it funny. As the interviewer asks Basse the same question, a soft toy lands on his desk. Basse jumps so high that for a split second he is completely aloft in the air before falling onto the floor. Other videos show Basse being teased by his colleagues—they throw balls of paper onto his desk, tap him on the back when he's not looking, even attach party horns to his chair so that they go off when he sits down. Each and every time, he screams, jumps and sometimes throws a punch.

There were certainly aspects of Basse's character that fit the Jumping Frenchman condition. Here he was, a source of entertainment among his friends, who had identified him as a Jumper and who had begun to startle him frequently, their laughter seemingly reinforcing his behavior.

I immediately got in touch with him. I explained that I thought he might be a modern-day Jumping Frenchman and wondered whether we could meet for a coffee. Which is how I came to be in Arendal.

When I arrive at Basse's office, it is late in the working day and most people have gone home. We sit opposite each other in a small, floor-to-ceiling glass-walled office. Basse tells me how it all started.

"It was that box prank that they videoed—once everyone saw how easily scared I was, they started doing it again and again. Now it happens all through the day."

He points to his desk. It is surrounded by high walls, near the front door. "I have to concentrate really hard while I'm working, so they can easily creep up behind me and make me jump. They do it all the time."

He recounts these tales with a huge smile on his face, and laughs out loud as he thinks back to the times he's jumped the most.

"The worst was in Amsterdam when I went to one of those dungeon experiences, where they take you around and scare you. I jumped so much that I threw up and had to be carried out."

He laughs again and shakes his head. "It's hard when it happens, it makes me shaky, but I also see the humor in it. Most of the time I just think to myself, 'What an idiot you are!'"

I wonder whether any of his family have the same overactive

startle response. "No," he says. "I have one brother and two sisters and they're not like this at all."

"Do you think it got worse once the guys at work cottoned on?" I ask.

"Absolutely," says Basse. "It's definitely become worse. Now I'm forever on high alert—waiting for the next time they're going to do it. I understand why they do it: it's funny. I normally don't mind, but sometimes when I'm really busy I ask them to be kind to me." He pauses. "Actually, I don't even need other people around to make me jump anymore."

"How's that?"

"I often make myself jump." He points to his collar. "Sometimes, just catching a glance of my own collar out of the corner of my eye will make me scream and jump."

Word of Basse has spread around the town. Now, whenever he goes out in public, people make him jump. "Everyone knows me, it happens everywhere," he says. "They'll do it when I'm out at the shops. Sometimes I just say to my wife, 'You go get the food, I can't face it!'"

When Basse is in a restaurant he has to sit in the corner. "That way, none of the waiters will need to tap me on the shoulder."

"Does it happen even when you're expecting it?" I say. "Like if I suddenly threw my arms up in the air?" I make the action after a short pause.

"Arghhh!" Basse's whole body instantly flies out of his chair, his feet kick the floor and send the chair backward into the glass. He thrashes his arms about as he lets out a huge, deep scream. I thought I'd given him enough time to anticipate my sudden movement, but I can see his heart pounding though his sweater as he catches his breath. He looks, just for an instant, exactly like the old lumberjack in the La-Z-Boy chair. And then his face quickly crinkles into laughter. "Oh my, and here I thought I was safe with you!"

AS I GET READY to leave the office, I remember to ask Basse whether he is okay with me using his full name in my book, or whether he wishes to remain anonymous.

"It's okay to use it," he says. "Except Basse is just my nickname."

"Oh yeah?"

"My real name is Hans Christian."

My heart does a little skip of its own. "So your full name is Hans Christian Andersen?"

"Yes." He laughs again. "I'm like a little fairy tale."

It seemed oddly appropriate.

Hans Christian Andersen was famous for telling wonderful tales of extraordinary beings whose actions would teach us something important about ourselves. It seemed like the ideal metaphor for what I had been trying to do all year.

Some scientists would argue that focusing on single people and the stories of their lives is far too subjective a way to teach us anything about the brain. I disagree. True, science prides itself on explaining the parts of our life that can be measured and tested. Objectivity is, rightly so, the backbone of science. But I'd argue that subjectivity is its flesh and blood. Each is necessary, but not sufficient alone. Alexander Luria called this individual portraiture "romantic science," a term that I would like to borrow. Let's inject a little more romance into the study of the brain—it may be the only way we will ever form a complete picture of what it has to offer.

I hope you've learned a little something about your own brain from the stories in this book. And when I say your brain, I really mean you. Because all too often we think about our brains as being somehow separate from ourselves. This is wrong. Waking up in the morning, feeling love for our children, searching for the answer to a desperately difficult problem—all of the things that make us who we are—are just functions of the activity whizzing through the squishy substance in our skull. All of our values, our emotions, our ideas are not, as Descartes would have it, floating around immaterial, they are all rooted in biology. Despite having worked with neuroscientists for my whole adult life, I have never fully appreciated that until now. It took seeing with my own eyes just how incredibly strange one's life can get when that activity is misplaced for me to truly understand that my brain and I are not two different things. We are our brains.

We don't have any adequate explanations yet for why our brain knows about itself. We so often eat whole meals without ever really tasting, arrive home without ever considering a direction, perhaps even spend whole days without really giving any thought to what we are doing. Why doesn't the brain just get on with the business of eating, fighting and procreating without the "me" that pops up to keep stock? Even with the advent of higher-resolution scanners, genetic manipulation and top-of-the-range medical technology, it's not a question we'll likely answer anytime soon. Our inability to understand our own minds is the price we pay for the ability to question them in the first place. Back in that first lesson with Clive, I was told by my professor that "if the brain were so simple that we could understand it, we would be so simple that we couldn't."

It goes without saying that we should relish the lives that it creates—particularly those that aren't "normal." The people who feature in this book are extraordinary, but my hope is that you have marveled at their humanity, rather than their eccentricity, that you have been surprised at the things we have in common, rather than the ways we are different. They have taught me that we each have an extraordinary brain. We may not have a memory as good as Bob's, but we can all reach back into our pasts and furnish our minds with millions of special moments. We may not hear music that doesn't exist or see colorful auras floating in the air, but we do hallucinate—our entire reality relies on it. We may never feel another person's pain as acutely as Joel, but thanks to our mirror neurons we do, indeed, feel it.

We all possess a remarkable feat of neural engineering that gives us intense feelings of love, that makes others laugh, that produces an unpredictable life that is utterly unique. It gives us the ability to remember an infinite amount of knowledge, to create an idea that has never been considered, to find an answer in the beating of our hearts. Our brain is a mystery that has not yet revealed the extent of the unimaginable lands it is capable of producing. And when it does, I think that will be the most romantic story of all.

Acknowledgments

I'd like to start by thanking Bob, Sharon, Rubén, Tommy, Shillo, Sylvia, Matar, Louise, Graham, Joel, Basse and all of their family and friends—the wonderful people who welcomed me into their homes, work and lives to let me tell their extraordinary stories. I am incredibly grateful to you all.

I would also like to express my greatest thanks to all of the scientists who gave up their time to talk to me about their work to ensure that it was depicted correctly in these pages.

Next, to my amazing editors: Georgina, Kate and Denise— thank you for your unceasing patience, guidance and incredible insight. You have been an absolute pleasure to work with at all times. On that same note, I'd also like to thank Cat, Jessica, Tiffany and Michael, whose editorial advice has lifted me on numerous occasions, and whose friendships I treasure.

To Max, my agent: I am so glad to be a part of the Brockman "family"—thank you for inviting me into it.

I also owe a considerable amount of gratitude to everyone at

New Scientist who helped me develop as a journalist and an editor. This project would not exist without you. Jeremy, you get a special mention: thank you for taking a chance on me all those years ago, even though you thought I was—what was it?—"completely unqualified for the job!"

I must also mention the rest of my friends, who willingly provide a nonjudgmental ear and a glass of wine at the drop of a hat—especially Be, Emily, Fatema and Sarah.

Although he is sadly no longer with us, I'd like to take a second to express my unwavering respect and love for the late Oliver Sacks, whose writing has inspired me throughout my life. I only got to speak to him in person once, but it was—of course—the most wonderful conversation I have ever had.

Finally, to my family: particularly my dad and my sisters, whose constant support and love has made it possible for me to embark on this adventure. I love you all so much. I have dedicated this book to Mum—I think she would have liked that— but there's a part of you all within it.

Last but not least, Alex. Thank you for your enduring love and infinite patience and encouragement, particularly over the past two years. I will be forever thankful for the fish-finger sandwiches that brought us together.

Notes and Sources

INTRODUCTION

1. The Edwin Smith Surgical Papyrus, Case 1 (1, 1-12). Translation by James P. Allen of the Metropolitan Museum of Art in New York.
2. Clarke, E., and O'Malley, C. D., "The Human Brain and Spinal Cord," *American Journal of Medical Sciences*, 17, 1968, pp. 467–69.
3. Caron, L., "Thomas Willis, the Restoration and the First Works of Neurology," *Medical History*, 59(4), 2015, pp. 525–53.
4. Ancient Greek and Roman physicians believed that four humors flowed throughout the brain and body. These were black bile, yellow bile, blood and phlegm. Hippocrates taught that an excess or deficit of any humor would result in ill health—an idea that was commonly held for centuries among European physicians.
5. Jay, Mike, *This Way Madness Lies: The Asylum and Beyond*, Thames & Hudson, 2016.
6. Sacks, Oliver, *The Man Who Mistook His Wife for a Hat*, Touchstone, 1985.

BOB

1. Corkin, Suzanne, *Permanent Present Tense: The Man with No Memory, and What He Taught the World*, Penguin, 2013.

2. Milner, B., et al., "Further Analysis of the Hippocampal Amnesic Syndrome: 14-Year Follow-Up Study of H.M.," *Neuropsychologia*, 6, 1968, pp. 215–34.

3. From Suzanne Corkin's account of her time with H.M.: "Henry Molaison: The Incredible Story of the Man with No Memory," *The Telegraph*, May 10, 2013.

4. Buñuel, Luis, *My Last Breath*, Vintage Digital, 2011, p. 121.

5. If you'd like to find out more about Solomon Shereshevsky and his fascinating memory, see: Luria, Alexander, *The Mind of a Mnemonist: A Little Book about a Vast Memory*, Harvard University Press, 1987.

6. McGaugh, J. L., et al., "A Case of Unusual Autobiographical Remembering," *Neurocase*, 12, 2006, pp. 35–49.

7. Foer, Joshua, *Moonwalking with Einstein*, Penguin Books, 2011.

8. Maguire, E., "Routes to Remembering: The Brains behind Superior Memory," *Nature Neuroscience*, 6(1), 2002, pp. 90–95.

9. McGaugh, J. L., et al., "A Case of Unusual Autobiographical Remembering."

10. Penfield, W., and Perot, P., "The Brain's Record of Auditory and Visual Experience: A Final Summary and Discussion," *Brain*, 86(4), 1963, pp. 595-696.

11. For a wonderful description of memory, please see Clare Wilson's feature: "What Does a Memory in My Brain Look Like?", *New Scientist*, 3049, November 28, 2015.

12. James, William, *Text-book of Psychology*, Macmillan, 1892.

13. Akers, K. G., et al., "Hippocampal Neurogenesis Regulates Forgetting during Adulthood and Infancy," *Science*, 344(6184), 2014, pp. 598–602.

14. Chris's story is described by Elizabeth Loftus in *Implicit Memory and Metacognition*, ed. Lynne Reder, Psychology Press, 1996.

15. You can find out more about Holland College and the Golden Knights here: www.cbsnews.com/news/a-60-minutes-story-you-will-never-forget.

16. LePort, A. K., et al., "Highly Superior Autobiographical Memory: Quality and Quantity of Retention over Time," *Frontiers in Psychology*, 6, 2016, p. 2017.

SHARON

1. Iaria, G., et al., "Developmental Topographical Disorientation: Case One," *Neuropsychologia*, 47(1), 2009, pp. 30–40.

2.   Ibid.

3.   Maguire, E. A., et al., "Navigation-Related Structural Change in the Hippocampi of Taxi Drivers," *PNAS*, 97(8), 2000, pp. 4398–403.

4.   Woollett, K., and Maguire, E. A., "Acquiring 'the Knowledge' of London's Layout Drives Structural Brain Changes," *Current Biology*, 21(24), 2011, pp. 2109–14.

5.   O'Keefe, J., "A Review of the Hippocampal Place Cells," *Progress in Neurobiology*, 13(4), 1979, pp. 419–39.

6.   Hafting, T., et al., "Microstructure of a Spatial Map in the Entorhinal Cortex," *Nature*, 436, 2005, pp. 801–6.

7.   "Geraldine Largay's Wrong Turn: Death on The Appalachian Trail," *New York Times*, May 26, 2016.

8.   "Use or Lose Our Navigational Skills," *Nature*, March 31, 2016.

9.   Woollett, K., et al., "Talent in the Taxi: A Model System for Exploring Expertise," *Philosophical Transactions of the Royal Society B*, 364, 2009, pp. 1407–16.

10. Sharon says she has recently come across two other people with developmental disorientation disorder who spin around to reset their mental maps. Both contacted her after hearing her describe the disorder on a podcast. One woman, like Sharon, had discovered the spinning technique as a child and has used it ever since.

11. Barclay, S. F., et al., "Familial Aggregation in Developmental Topographical Disorientation (DTD)," *Cognitive Neuropsychology*, 6, 2016, pp. 1–10.

RUBÉN

1.   Haraldsson, Eriendur, and Gissurarson, Loftur, *Indridi Indridason: The Icelandic Physical Medium*, White Crow Productions, 2015.

2.   Gissurarson, L. R., and Gunnarsson, A., "An Experiment with the Alleged Human Aura," *Journal of the American Society for Psychical Research*, 91, 1997, pp. 33–49.

3.   A translation of Sachs's dissertation appears in the following paper: Jewanski, J., et al., "A Colourful Albino: The First Documented Case of Synaesthesia, by Georg Tobias Ludwig Sachs in 1812," *Journal of the History of the Neurosciences*, 18(3), 2009, pp. 293–303.

4.   Nabokov, Vladimir, *Speak, Memory: An Autobiography Revisited*, Penguin Modern Classics, 2012, pp. 23–25.

5.   Bor, D., et al., "Adults Can Be Trained to Acquire Synesthetic Experiences," *Nature Scientific Reports*, 4, 2014, p. 7089.

6. Ramachandran explores this subject in greater depth in: Ramachandran, V. S., *The Tell-Tale Brain: Unlocking the Mystery of Human Nature*, Cornerstone Digital, 2012.

7. Atkinson, J., et al., "Synesthesia for Manual Alphabet Letters and Numeral Signs in Second-Language Users of Signed Languages," *Neurocase*, 22(4), 2016, pp. 379–86.

8. Chun, C. A., and Hupe, J.-M., "Mirror-Touch and Ticker Tape Experiences in Synesthesia," *Frontiers in Psychology*, 4, 2013, p. 776.

9. Nielsen, J., et al., "Synaesthesia and Sexuality: The Influence of Synaesthetic Perceptions on Sexual Experience," *Frontiers in Psychology*, 4, 2013, p. 751.

10. Kayser, D. N., et al., "Red and Romantic Behavior in Men Viewing Women," *European Journal of Social Psychology*, 40(6), 2010, pp. 901–8.

11. Attrill, M. J., et al., "Red Shirt Colour Is Associated with Long-Term Team Success in English Football," *Journal of Sports Sciences*, 26(6), 2008, pp. 577–82.

12. Hill, R. A., and Barton, R. A., "Red Enhances Human Performance in Contests," *Nature*, 435, 2005, p. 293.

13. I have written more about the evolutionary rules of attraction in: "Darwinian Dating: Baby, I'm Your Natural Selection," *New Scientist*, 2799, February 12, 2011.

14. For this, I have relied heavily upon the excellent explanations given by Adam Rogers, in: "The Science of Why No One Agrees on the Color of This Dress," *Wired*, February 26, 2015.

15. Milán, E. G., et al., "Auras in Mysticism and Synaesthesia: A Comparison," *Consciousness and Cognition*, 21, 2011, pp. 258–68.

16. Ramachandran, V. S., and Hubbard, E. M., "Psychophysical Investigations into the Neural Basis of Synaesthesia," *Proceedings of the Royal Society B*, 268, 2001, pp. 979–83.

TOMMY

1. Burns, J. M., and Swerdlow, R. H., "Right Orbitofrontal Tumor with Pedophilia Symptom and Constructional Apraxia Sign," *Archives of Neurology*, 60, 2003, pp. 437–40.

2. For more details of all of these studies, please see: Segal, Nancy, *Born Together—Reared Apart: The Landmark Minnesota Twin Study*, Harvard University Press, 2012.

3. Segal, N., et al., "Unrelated Look-Alikes: Replicated Study of Personality Similarity and Qualitative Findings on Social Relatedness," *Personality and Individual Differences*, 55(2), 2013, pp. 169–74.
4. Gatz, M., et al., "Importance of Shared Genes and Shared Environments for Symptoms of Depression in Older Adults," *Journal of Abnormal Psychology*, 101(4), 1992, pp. 701–8.
5. Kosslyn, Stephen, and Miller, G. Wayne, *Top Brain, Bottom Brain: Surprising Insights into How You Think*, Simon & Schuster, 2013.
6. Some of Tommy's artwork has been published online at: www .tommymchugh.co.uk.
7. Flaherty explores her own overwhelming urge to write and that of others in: Flaherty, Alice, *The Midnight Disease: The Drive to Write, Writer's Block, and the Creative Brain*, Mariner Books, 2005.
8. Woollacott, I. O., et al., "Compulsive Versifying after Treatment of Transient Epileptic Amnesia," *Neurocase*, 21(5), 2015, pp. 548–53.
9. Woolley, A. W., et al., "Using Brain-Based Measures to Compose Teams: How Individual Capabilities and Team Collaboration Strategies Jointly Shape Performance," *Social Neuroscience*, 2(2), 2007, pp. 96–105.

SYLVIA

1. Jardri, Renaud, et al., eds., *The Neuroscience of Hallucinations*, Springer, 2013.
2. Sacks, Oliver, *Hallucinations*, Picador, 2012.
3. Part of this chapter has been adapted from a feature I wrote: "Making Things Up," *New Scientist*, 3098, November 5, 2016.
4. Ffytche, D. H., et al., "The Anatomy of Conscious Vision: An fMRI Study of Visual Hallucinations," *Nature Neuroscience*, 1(8), 1998, pp. 738–42.
5. Charles Bonnet, 1760, as quoted by Oliver Sacks, TED talk: "What Hallucination Reveals about Our Minds," 2009.
6. Rosenhan, D. L., "On Being Sane in Insane Places," *Science*, 179, 1973, pp. 250–58.
7. McGrath, J. J., et al., "Psychotic Experiences in the General Population," *JAMA Psychiatry*, 72(2), 2015, pp. 697–705.
8. Wackermann, J., et al., "Ganzfeld-Induced Hallucinatory Experience, Its Phenomenology and Cerebral Electrophysiology," *Cortex*, 44, 2008, pp. 1364–78.

9. Frith, Chris, *Making Up the Mind: How the Brain Creates Our Mental World*, Wiley-Blackwell, 2007, p. 111.

10. Kumar, S., et al., "A Brain Basis for Musical Hallucinations," *Cortex*, 52(100), 2014, pp. 86–97.

11. Daniel, C., and Mason, O. J., "Predicting Psychotic-Like Experiences during Sensory Deprivation," *BioMed Research International*, 2015, 439379.

MATAR

1. Woodwood, Ian, *The Werewolf Delusion*, Paddington Press, 1979, p. 48.

2. As recounted by Russell Hope Robbins in *The Encyclopaedia of Witchcraft and Demonology*, Springer Books, 1967, p. 234.

3. Moselhy, H. F., "Lycanthropy, Mythology and Medicine," *Irish Journal of Psychological Medicine*, 11(4), 1994, pp. 168–70.

4. Keck, P. E., et al., "Lycanthropy: Alive and Well in the Twentieth Century," *Psychological Medicine*, 18(1), 1988, pp. 113–20.

5. Toyoshima, M., et al., "Analysis of Induced Pluripotent Stem Cells Carrying 22q11.2 Deletion," *Translational Psychiatry*, 6, 2016, e934.

6. Frith, C. D., et al., "Abnormalities in the Awareness and Control of Action," *Philosophical Transactions of the Royal Society B*, 355, 2000, pp. 1771–88.

7. Lemaitre, A.-L., et al., "Individuals with Pronounced Schizotypal Traits Are Particularly Successful in Tickling Themselves," *Consciousness and Cognition*, 41, 2016, pp. 64–71.

8. Large, M., et al., "Homicide Due to Mental Disorder in England and Wales Over 50 Years," *British Journal of Psychiatry*, 193(2), 2008, pp. 130–33.

9. The science writer Mo Costandi has written a wonderful description of Penfield's life and work in his blog: "Wilder Penfield, Neural Cartographer," www.neurophilosophy.wordpress.com, August 27, 2008.

10. McGeoch, P. D., et al., "Xenomelia: A New Right Parietal Lobe Syndrome," *Journal of Neurology, Neurosurgery and Psychiatry*, 82(12), 2011, pp. 1314–19.

11. Case, L. K., et al., "Altered White Matter and Sensory Response to Bodily Sensation in Female-to-Male Transgender Individuals," *Archives of Sexual Behavior*, pp. 1–15.

LOUISE

1. *Amiel's Journal: The Journal Intime of Henri-Frédéric Amiel*, trans. Mrs. Humphrey Ward, A. L. Burt Company, 1900.
2. As recalled by Gerd Woll, senior curator at the Munch Museum, in Arthur Lubow's *Edvard Munch: Beyond The Scream*, Smithsonian, 2006.
3. As translated by the Munch Museum, www.emunch.no.
4. http://www.dpselfhelp.com/forum.
5. Couto, B., et al., "The Man Who Feels Two Hearts: The Different Pathways of Interoception," *Social Cognitive and Affective Neuroscience*, 9(9), 2014, pp. 1253–60.
6. Damasio, Antonio, *Descartes' Error: Emotion, Reason and the Human Brain*, Vintage Digital, 2008.
7. You can hear more from Damasio on this subject here: www.scientificamerican.com/article/feeling-our-emotions.
8. Medford, N., et al., "Emotional Experience and Awareness of Self: Functional MRI Studies of Depersonalization Disorder," *Frontiers in Psychology*, 7(432), 2016, pp. 1–15.
9. Medford, N., "Emotion and the Unreal Self: Depersonalization Disorder and De-affectualization," *Emotion Review*, 4(2), 2012, pp. 139–44.
10. Khalsa, S. S., et al., "Interoceptive Awareness in Experienced Meditators," *Psychophysiology*, 45(4), 2007, pp. 671–77
11. Ainley, V., et al., "Looking into Myself: Changes in Interoceptive Sensitivity during Mirror Self-Observation," *Psychophysiology*, 49(11), 2012, pp. 1504–8.

GRAHAM

1. Pearn, J., and Gardner-Thorpe, C., "Jules Cotard (1840–1889): His Life and the Unique Syndrome which Bears His Name," *Neurology*, 58, 2002, pp. 1400-3.
2. Ibid.
3. Cotard, J.-M., "Du Délire des Négations," *Archives de Neurologie*, 4, 1882, pp. 152–70. (Thank you to Jennifer Halpern, who translated the chapter from French to English for me.)
4. Pearn and Gardner-Thorpe, "Jules Cotard."
5. Clarke, Basil, *Mental Disorder in Earlier Britain: Exploratory Studies*, University of Wales Press, 1975.

6. Lemnius, Levinus, *The Touchstone of Complexions*, Marshe, 1581, title page.

7. Ibid.

8. Ibid., p. 152.

9. Owen, A. M., et al., "Detecting Awareness in the Vegetative State," *Science*, 313, 2006, p. 1402.

10. Yu, F., et al., "A New Case of Complete Primary Cerebellar Agenesis: Clinical and Imaging Findings in a Living Patient," *Brain*, 138(6), 2015, e353.

11. Kelly Servick, "A Magnetic Trick to Define Consciousness," *Wired*, August 15, 2013.

12. Casali, A. G., et al., "A Theoretically Based Index of Consciousness Independent of Sensory Processing and Behavior," *Science Translational Medicine*, 5(198), 2013.

13. Koubeissi, M. Z., et al., "Electrical Stimulation of a Small Brain Area Reversibly Disrupts Consciousness," *Epilepsy & Behavior*, 37, 2014, pp. 32–35.

14. Charland-Verville, V., et al., "Brain Dead Yet Mind Alive: A Positron Emission Tomography Case Study of Brain Metabolism in Cotard's Syndrome," *Cortex*, 49(7), 2013, pp. 1997–99.

15. Lindén, T., and Helldén, A., "Cotard's Syndrome as an Adverse Effect of Acyclovir Treatment in Renal Failure," *Journal of the Neurological Sciences*, 333(1), 2013, e650.

16. As referred to by Hans Forstl and Barbara Beats in "Charles Bonnet's Description of Cotard's Delusion and Reduplicative Paramnesia in an Elderly Patient (1788)," *British Journal of Psychiatry*, 160, 1992, pp. 416–18.

17. Ryle, Gilbert, *The Concept of Mind*, Peregrine, 1949, pp. 186–89.

JOEL

1. di Pellegrino, G., et al., "Understanding Motor Events: A Neurophysiological Study," *Experimental Brain Research*, 91(1), 1992, pp. 176–80.

2. Perry, A., et al., "Mirroring in the Human Brain: Deciphering the Spatial-Temporal Patterns of the Human Mirror Neuron System," *Cerebral Cortex*, 2017, pp. 1–10.

3. Blakemore, S.-J., et al., "Somatosensory Activations during the Observation of Touch and a Case of Vision-Touch Synaesthesia," *Brain*, 128(7), 2005, pp. 1571–83.

4. Banissy, M. J., et al., "Superior Facial Expression, but Not Identity Recognition, in Mirror-Touch Synaesthesia," *Journal of Neuroscience*, 31(5), 2011, pp. 1820–24.

5. Ward, J., and Banissy, M. J., "Explaining Mirror-Touch Synesthesia," *Cognitive Neuroscience*, 6(2–3), 2015, pp. 118–33.

6. Santiesteban, I., et al., "Mirror-Touch Synaesthesia: Difficulties Inhibiting the Other," *Cortex*, 71, 2015, pp. 116–21.

7. Kramer, A. D. I., et al., "Experimental Evidence of Massive-Scale Emotional Contagion through Social Networks," *PNAS*, 111(24), 2014, pp. 8788–90.

8. Meffert, H., et al., "Reduced Spontaneous but Relatively Normal Deliberate Vicarious Representations in Psychopathy," *Brain*, 136(8), 2013, pp. 2550–62.

9. Singer, T., and Klimecki, O. M., "Empathy and Compassion," *Current Biology*, 24(18), 2014, R875–78.

CONCLUSION

1. Beard, G., "Remarks upon Jumpers or Jumping Frenchmen," *Journal of Nervous Mental Disorders*, 5, 1878, p. 526.

2. Beard, G., "Experiments with the Jumpers of Maine," *Popular Science Monthly*, 18, 1880, pp. 170–78.

3. Saint-Hilaire, M.-H., et al., "Jumping Frenchmen of Maine," *Neurology*, 36, 1986, p. 1269.

4. "The most easily scared guy in the world?", December 14, 2012, https://www.youtube.com/watch?v=WfQ4t2E7iAU.

Index